The Organic Codes

The genetic code appeared on Earth with the first cells. The codes of cultural evolution arrived almost 4 billion years later. These are the only codes that are recognised by modern biology. In this book, however, Marcello Barbieri explains that there are many more organic codes in nature, and their appearance not only took place throughout the history of life but marked the major steps of that history. A code establishes a correspondence between two independent "worlds", and the codemaker is a third party between those "worlds". Therefore the cell can be thought of as a trinity of genotype, phenotype and ribotype. The ancestral ribotypes were the agents which gave rise to the first cells.

The book goes on to explain how organic codes and organic memories can be used to shed new light on the problems biologists encounter in cell signalling, epigenesis and embryonic development. A mathematical model is presented to show how embryos can increase their own complexity by the use of a code and a memory.

This fascinating book will be of interest to all biologists and anyone with an interest in the origins and the evolution of life on Earth.

MARCELLO BARBIERI has conducted research on embryonic development and ribosome crystallisation at the Medical Research Council in Cambridge, UK, the National Institutes of Health in Bethesda, USA, and the Max-Planck-Institut für Molekulare Genetik in Berlin, Germany. He has published books on embryology and evolution, and has taught biophysics, molecular embryology and theoretical biology respectively at the Universities of Bologna, Sassari and Turin, Italy. Since 1992 he has taught embryology and conducted research in theoretical biology at the University of Ferrara, Italy.

THE ORGANIC CODES
An introduction to semantic biology

Marcello Barbieri
Università di Ferrara

CAMBRIDGE
UNIVERSITY PRESS

CAMBRIDGE
UNIVERSITY PRESS

University Printing House, Cambridge CB2 8BS, United Kingdom

Cambridge University Press is part of the University of Cambridge.

It furthers the University's mission by disseminating knowledge in the pursuit of education, learning and research at the highest international levels of excellence.

www.cambridge.org
Information on this title: www.cambridge.org/9780521531009

© Cambridge University Press 2003

First published 2003

A catalogue record for this publication is available from the British Library

Library of Congress Cataloguing in Publication data

Barbieri, Marcello.
   The organic codes: an introduction to semantic biology /
   Marcello Barbieri; drawings by Elisa Padovani.
     p. cm.
   Includes bibliographical references and index.
   ISBN 0 521 82414 1 – ISBN 0 521 53100 4 (pb.)
1. Biology – Philosophy. 2. Semantics (Philosophy) 3. Evolution (Biology)
4. Developmental biology. 5. Epigenesis – Mathematical models. I. Title.

QH331.B247 2002
570´.1 – dc21   2002073767

ISBN 978-0-521-53100-9 Paperback

CONTENTS

FOREWORD

Most scientific publications deal with problems that can be explained in a straightforward manner and with solutions that can be evaluated as a matter of routine. But scientific progress often occurs when somebody tries to reformulate the problem, or to suggest a different kind of solution. When that happens, it may be necessary to dwell as much upon the questions as upon the answers, and to show how a novel approach might give further significant results.

Barbieri finds that biology has been able to deal with information and with structure, but not with the connection between them. Something has been left out, and that is meaning. Semantics is the branch of logic that deals with meaning: hence the term "semantic biology". Meaning is a difficult concept to analyse, even though we find it in everything we read or listen to, including imaginative literature. To understand a poem one needs all sorts of background information. Poetry is rich in literary allusions, so just knowing the words will not do. Meaning is largely a matter of context, and that makes it hard to pin down.

The contextuality of meaning may be called a "principle", for it is neither a brute fact nor a law of nature. But exactly what is meant by a principle is hard to specify. We can give some familiar examples of course. In ecology there is the well-known "competitive exclusion principle", which explains why organisms occupying exactly the same niche cannot coexist for more than a brief period of time. In logic we all use, whether we know it or not, the "principle of contradiction", which states that two propositions that really contradict each other cannot both be true. And since, by implication, at least one of them must be false, we justify the kind of hypothetico-deductive scientific method that Barbieri (an admirer of Popper) endorses. Principles are very important in science, more important than may seem obvious. Usually we adopt them implicitly, without giving them much thought. Principles are perhaps the most important components of Barbieri's theoretical, or perhaps better, metatheoretical, system. One might even say that such principles are what the book is really all about.

Barbieri enunciates four general principles, all of which relate to the problems of development. He begins by considering epigenesis, and redefines it as the property of a system to increase its own complexity. He goes so far as to make the capacity for attaining such convergent complexity both a fundamental principle and a defining property of life itself. One might question that, but his definition is at least as good as any of the ones that are quoted in the Appendix. The second principle tells us that achieving such convergent complexity amounts to reconstructing a structure from incomplete information. That in turn provides a new definition of "epigenesis". Then we get a third principle, according to which organic epigenesis requires organic memories. Here "memory" is a technical term indicating that there has to be some repository of information. And as a final principle, such epigenesis requires organic codes. Indeed codes and memories exist only because they are necessary for producing epigenetic systems.

Barbieri is a scientist, not a philosopher. He justifies his ideas on the basis of their ability to make sense out of the material universe. This he accomplishes by means of four "models", as he calls them. Why "models" rather than "theories"? Evidently because they serve to illustrate the principles. Of course it really matters whether the particular interpretations are correct. But the point of the book could be made just as well if the hypotheses being discussed were modified in some respects. The more basic message is not the examples as such, but rather the kind of theory that might be expected to emerge out of a semantic approach to biology. Let us have a brief look at these models from that perspective.

First, Barbieri presents a theory about the origin of life. Extant organisms possess both genotype, in the form of DNA molecules, and phenotype, in the form of proteins, cells, and other products of epigenesis. Previous scenarios treated proteins or DNA as coming first. Both of these alternatives ran into difficulties because the one cannot exist without the other. For that very reason there must have been something additional to genotype and phenotype, which he calls the ribotype. It is RNA that bridges the gap between genotype and phenotype, and it does so by endowing the system with meaning. Cells contain all three. Those who want to define life as either as genes or as gene products will find no comfort in this view of it.

The second model illustrates the point that more than one kind of memory can be responsible for the reconstruction from incomplete information that takes place during the (epigenetic) formation of an organism. Barbieri proposes that two kinds of memory are in fact responsible for the development of multicellular animals – one for the earlier stages, the other for the later ones. He shows how the existence of these two kinds of memory might account for the pattern of macroevolution, notably the Cambrian explosion.

The third model is an application of similar considerations to mental development, especially with respect to language. One kind of organic memory accounts for the acquisition of the capacities that appear early in the ontogeny of language, then a second takes over. Again, codes are absolutely indispensable, and the emergence of new ones has been a key innovation in the history of both life and mind.

And finally, the semantic theory applies to culture as well. Cultures are like species, insofar as they are supraorganismal wholes, and real concrete things. There are codes in both life and culture, and both life and culture have evolved through natural selection and natural conventions. In culture we find something analogous to genotypes, though they depend upon an extrasomatic memory. We also find something analogous to phenotypes, such as artifacts. But, if we are to extend Barbieri's basic vision of organised beings to culture, there is also something more. Consider a village with its buildings. Is it blueprints that explain the existence of buildings, or buildings that explain the existence of blueprints? Barbieri suggests that we might ask more edifying questions.

Barbieri's most ambitious claim is that life evolves through natural conventions as well as natural selection. The importance of such conventions as major evolutionary innovations becomes increasingly obvious as he discusses one example after another. Yet let us not get carried away. There is nothing here that portends the fall of Darwinism or its replacement by an alternative paradigm. The book is, after all, concerned with the fundamental principles of development, and with how they relate to the grand picture of evolution. It belongs to the mainstream of biological thought, and finds its proper place among the works of Karl Ernst von Baer, Charles Darwin, and August Weismann.

October 2001   *Michael T. Ghiselin*

DEDICATION

To   *Karl Popper*
      René Thom
      Heinz-Günter Wittmann
      and Elmar Zeitler

This book is an extension of *The Semantic Theory of Evolution* (1985) and is dedicated, with affection, to the four men who encouraged my long journey toward that view of life.

Karl Popper has been my most important spiritual referee, and his pronouncement, in a private letter, that the semantic theory of life is *"revolutionary"* gave me the strength to persevere.

René Thom has been the *deus ex machina* who actually engineered the publication of *The Semantic Theory* and gave it an impressive *imprimatur* by writing its preface.

Heinz-Günther Wittmann and **Elmar Zeitler** allowed me to perform the experimental research which led me first to the concept of *ribotype* and then to the idea of *evolution by natural conventions*.

It is from these good men that I learned what it takes to devote one's life to an idea, even if all seems to be destined to another generation of students. Which is what really matters, in the end, because a new idea is all the more beautiful the greater is its power to convince one that it really belongs to the future.

March 2002   *Marcello Barbieri*

ACKNOWLEDGEMENTS

I was greatly encouraged by the comments to the first version of this book made by Robert Aunger, Noam Chomsky, Eva Jablonka, Kalevi Kull, Richard Strohman and Emile Zuckerkandl. And I have vastly profited from the suggestions sent by David Abel, James Barham, Jack Cohen, Michael Ghiselin, Chris Ottolenghi and Pietro Ramellini, and from the corrections supplied by James Barham, Richard Gordon and Romeu Guimarães. Michael Ghiselin has written a Foreword that makes an illuminating synthesis of the theoretical structure of the book, which suddenly comes to life, and for that kind of help I simply cannot find the right words. Jack Cohen has added an Afterword which puts some experimental flesh on the backbone of theory, thus ending off the book on a very encouraging note. Ward Cooper is "the" editor who adopted the book into the CUP family, where it has been carefully nursed by Carol Miller (production controller), Anna Hodson (copy editor) and Zoe Naylor (cover designer) all the way up to its present state. To all above colleagues and friends: thank you, I will never forget you.

July 2002   *Marcello Barbieri*

INTRODUCTION

There is a strange paradox in modern biology. On the one hand, new discoveries are made at such a high rate that our science of life appears full of surprises and in a constant state of flux. On the other hand, all new findings are apparently accommodated within a theoretical framework that remains remarkably stable. Present-day biology, in other words, seems to be in that phase of development that Thomas Kuhn referred to as *"normal science"*, a phase in which an endless stream of novelties is smoothly accounted for by an unchanging paradigm. And this is definitely not for want of alternatives. No efforts have been spared to provide different explanations of life, but none has withstood the test of time. What makes us feel good about our present paradigm (which many call *universal Darwinism*) is that only the truth – or something very near the truth – can resist so many assaults and outlive generations of critics. In such a situation, I find it almost embarrassing to suggest that our beloved paradigm is not as perfect as we like to think. But this is the message that is coming from nature, and I had better tell you straight away the reasons that lead to this conclusion. The main points are three: the existence of organic codes, a mathematical model of epigenesis and a new theory of the cell.

The organic codes

From time immemorial it has been thought that codes, or conventions, exist only in the mutable world of culture, while nature is governed by immutable laws. The discovery that a genetic code is at the very heart of life came therefore as a bolt from the blue. And people rushed to anaesthetise it. The genetic code was immediately declared a *frozen accident*, and the divide between nature and culture remained substantially intact. The existence of other organic codes is, in principle, as natural as that of the genetic code, but its implications are perhaps even more revolutionary. The genetic code appeared on Earth with the first cells, while the linguistic codes arrived almost 4 billion years later, with cultural evolution. These are the only codes that modern biology currently recognises, which is tantamount to saying that in 4 billion years no other code appeared on our planet. And if codes are relegated to the beginning and to the end of the history of life, we can safely say that 4 billion years of biological evolution went on with the sole mechanism of natural selection. In this book, however, we will see that there are many other organic codes in nature, and that they appeared not only throughout the history of life but marked the main steps of that history, the steps which brought about the great events of macroevolution. But if codes exist, they must have had origins and histories, and above all they must have had a specific mechanism. Languages evolved not only by chance mutations of letters in their words but also by changes in their grammatical rules, and the same would apply to living organisms. We must conclude, in short, that biological evolution was produced by two distinct mechanisms: *by natural selection and by natural conventions*.

From a logical point of view this is a straightforward conclusion, but unfortunately theory and practice do not always go hand in hand. The idea of evolution by natural conventions was proposed for the first time in 1985, in a book of mine entitled *The Semantic Theory of Evolution*, but it did not have any significant impact (even if I am pleased to say that in a private letter Karl Popper called it *"revolutionary"*). Regrettably, people do not seem to associate the existence of organic codes with a mechanism of natural conventions, as if one could exist without the other. Edward Trifonov, for example, has been campaigning in favour of sequence codes since 1988, and in 1996 William Calvin wrote a book entitled *The Cerebral Code*, but nobody called for anything different from natural selection. And there is a reason for that. The reason is that the word *code* has largely been used in a metaphorical sense, as have so many other words which have been borrowed by molecular biologists from everyday language. It is imperative, therefore, to realise that there are organic codes which are not metaphorical but real, and to this purpose we clearly need to prove their existence. A code is a correspondence between two independent worlds, and a real organic code requires molecules that perform two independent recognition processes. These are the codes' fingerprints, and it is they that we must look for and bring to light. In the genetic code these molecules are the transfer RNAs, but we will see that equivalent *adaptors* (the word that Francis Crick initially proposed for the tRNAs) exist in at least two other processes (signal transduction and splicing) and are expected to turn up in many other cases. And luckily this is beginning to happen. In the year 2000, for example, Gabius provided evidence for a *sugar code*, while Strahl, Allis, Turner and colleagues discovered a *histone code*. The more we learn about organic codes, in conclusion, the more they turn out to be every bit as real as the genetic code. Sooner or later, therefore, biologists will have to come to terms with the theoretical implications of this extraordinary experimental fact.

A mathematical model of epigenesis

Embryonic development was defined by Aristotle as an *epigenesis*, i.e. as a sequence of one genesis after another, a step-by-step generation of new structures. Today epigenesis is often referred to as an *increase of complexity*, but when we use this expression we should always add an important qualification. We should say that epigenesis is a *convergent* increase of complexity, in the sense that its outcome is neither random nor unexpected. This is what makes it so radically different from the *divergent* increase that takes place in evolution. The distinction between convergent and divergent phenomena is particularly relevant today that the study of complexity has become a research field in its own right. Many interesting ways of obtaining *"order out of chaos"* have been described and have found applications in various disciplines, but the expectation that they could apply to embryonic development has been an illusion. Embryos are not chaotic systems, and embryonic stages are not phase transitions.

To my knowledge, there is only one mathematical model which has described how a convergent increase of complexity can actually take place. I developed this model as a special case of the general problem of reconstructing structures from projections, a problem which arises in fields as diverse as radioastronomy, electron microscopy and computerised tomography. The mathematics of the reconstruction problem has been thoroughly investigated, and the minimum number of projections required for a complete reconstruction is prescribed by basic theorems. This allows us to give a precise formulation to a problem which may seem hopeless at first sight: the problem of reconstructing structures *from incomplete information*. We can legitimately say that we are performing this type of reconstruction when we work with a number of projections which is at least one order of magnitude less than the theoretical minimum, i.e. when we use 10% or less of the minimal information. What is interesting about this strange-looking problem is that *a reconstruction from incomplete information* is equivalent, to all practical purposes, to *a convergent increase of complexity*, and so it is a mathematical formulation of the problem of epigenesis (if the starting information is incomplete, the reconstruction must produce an increase of information and this is equivalent to an increase of complexity). Even more interesting is that the problem can actually be solved, as we will see in Chapter 3. And the beauty of the solution is that its logic can be grasped even without the mathematics (which will however be provided). The model employs an iterative procedure that performs in parallel two different reconstructions: one for the structure in question and one for its *reconstruction memory*. The key point is that the memory space turns out to have the surprising ability to provide new specific information about the examined structure, and such information can be transferred from the memory space to the structure space with appropriate codes, or conventions. The conclusion is that a convergent increase of complexity can be achieved if a reconstruction is performed with memories and codes. Which means, in biological terms, that epigenesis requires organic memories and organic codes. We come back, in this way, to the issue of the organic codes. And the message from mathematics is strong: there is no way that we are going to understand a phenomenon as large as embryonic development without organic codes and organic memories.

A new theory of the cell

The extraordinary thing about codes is that they require a new entity. In addition to energy and information they require *meaning*. For centuries, meaning has been regarded as a spiritual or a transcendental entity, but the very existence of the genetic code proves that it is as natural as information. And in fact we can define meaning with an operative procedure just as we do with any other natural entity. *Meaning is an object which is related to another object via a code.* The meaning of the word *apple*, for example, is the mental object of the fruit which is associated to the mental object of that word by the code of the English language (needless to say, the code of another language would associate a different mental object to the same word). The meaning of a combination of dots and dashes is a letter of the alphabet, in the Morse code. The meaning of a combination of three nucleotides is usually an amino acid, in the genetic code (from which it follows that the meaning of a gene is usually a protein).

We are well aware that it is man who gives meaning to mental objects – in the realm of the mind he is the *codemaker* – but this does not mean that a correspondence between two independent worlds must be the result of a conscious activity. The only logical necessity is that the codemaker is *an agent that is ontologically different* from the objects of the two worlds, because if it belonged to one of them the two worlds would no longer be independent. A code, in other words, requires three entities: two independent worlds and a codemaker which belongs to a third world (from a philosophical point of view this is equivalent to the triadic system proposed in semiotics by Charles Peirce). In the case of the genetic code, the codemaker is the ribonucleoprotein system of the cell, a system which operates as a true third party between genes and proteins. This is why I proposed, in 1981, that the cell is not a duality of genotype and phenotype but a trinity made of genotype, phenotype and *ribotype*. And I argued that the ribotype is a cell category that not only has the same ontological status as genotype and phenotype, but has a logical and a historical priority over them (hence the title of the paper: "The ribotype theory on the origin of life").

The fact that the ribotype is the codemaker of the genetic code leads necessarily to a change of our traditional view of the cell, but there is also another reason for this theoretical shift. The definitions of life that have been proposed in the last 200 years (for a compendium see the Appendix), have underlined a variety of presumed *essential* features (heredity, replication, metabolism, autonomy, homeostasis, autopoiesis, etc.), but none of them has ever mentioned *epigenesis* as a defining characteristic of life. The reason of course is that epigenesis has been associated with embryos, not with cells, and yet even in single cells the phenotype is always more complex than the genotype. This means that every cell has the ability to increase its own complexity, and so it really is an epigenetic system. We realise in this way that the mere presence of organic codes in every cell, starting from the genetic code, requires a theoretical framework where organic meaning is a necessary complement of organic information. And that is precisely what semantic biology is about. It is not a denial of our Darwinian paradigm. It is a genuine extension of it.

About this book

Chapters 1 and 2 are an introduction to the cell theory and to the theories of evolution at a level that may be regarded as undergraduate or thereabouts. Those who are not concerned with undergraduates may skip them and start with Chapter 3, but should not forget that semantic biology applies to all levels of the life sciences and is not just a section for specialists. Chapter 3 presents a model of epigenesis first in words and then in formulae, and even the biologists who are not devotees of mathematics can follow it from beginning to end. This is highly recommended because the idea that a structure can be reconstructed from incomplete information is still met with incredulity in our present educational system (if engineers and computer scientists insist that it can't be done, just tell them that embryos do it all the time). Chapter 4 is a biological sequel of Chapter 3 and makes a first excursion into the world of organic codes and organic memories. This is instrumental to the next three chapters which are dedicated to the main events of macroevolution: the origin of life (Chapter 5), the emergence of eukaryotic cells (Chapter 6) and the Cambrian explosion of animals (Chapter 7). These chapters allow us to revisit those great transitions and show how different they look like when organic codes are taken into account. Chapter 8 brings together the ideas of the previous chapters and presents a first outline of the framework of semantic biology. And in order to underline the logical structure of this framework, Chapter 9 makes a brief summary of it in eight propositions (four principles and four models).

The chapters of this book are arranged in a sequential order, but they are also largely autonomous and one can read them in any order. Everything in biology is linked to everything else, and it doesn't really matter where one starts from. What does matter is that whichever way we look at life today we realise that organic codes are there, that they have always been there, from the very beginning, and that it is about time we start taking notice.

1

THE MICROSCOPE AND THE CELL

The cell theory and the theory of evolution are the two pillars of modern biology, but only the latter seems to be the object of ongoing research and debates. The cell theory is generally regarded as a closed chapter, a glorious but settled issue in the history of science. The emphasis today is on cell experiments, not on cell theory, and there is no doubt that one of our greatest challenges is the experimental unravelling of the extraordinary complexity that has turned out to exist in the cellular world. At various stages of this book, however, we will see that the experimental results suggest new ideas, and at the end of the book it will be possible to combine them in a new model of the cell. This is because cells are not only what we see in a biological specimen through the lenses of a microscope, but also what we see through the lenses of a theory. The cell, after all, is a system, and understanding the logic of a system requires some theorising about it. And since this theorising has a long history behind it, let us begin by retracing the main steps of that intellectual journey. This chapter shows that the concept of the cell had to be imposed on us by the microscope because it was unthinkable in the world-view of classic philosophy. And after that intrusion, the concept has gradually changed and in so doing it has changed our entire approach to the problems of generation and embryonic development. But this historical journey is not without surprises, because it will take us toward an idea that all definitions of life of the last 200 years have consistently missed. The idea that epigenesis does not exist only in embryos but in every single cell. That the phenotype is always more complex that the genotype. That epigenesis is a defining characteristic of life.

The cell theory

The idea that all living creatures are made of cells has changed more than anything else our concept of life, and is still the foundation of modern biology. This great generalization was made possible by the invention of the microscope, but did not come suddenly. It has been the culmination of a collective research which lasted more than two hundred years, and in order to understand it we must be aware of the main problems that had to be solved.

Let us start with the microscope. Why do we need it? Why can't we see the cells with the naked eye? The answer is that the eye's retina itself is made of cells. Two objects can be seen apart only if their light rays fall on different cells of the retina, because if they strike the same cell the brain receives only one signal. More precisely, the brain can tell two objects apart only when their images on the retina have a distance between them of at least 150 μm (thousandths of a millimetre). The cells have average dimensions (10 μm) far smaller than that limit, and, even if an organism is stared at from a very close distance, their images overlap and they remain invisible. It is therefore necessary to enlarge those images *in order to increase their distance on the retina*, and that is where the microscope comes in.

Enlargements of 5 or 10 times can be obtained with a single lens (the so-called simple microscope) but are not enough for seeing the cells. Substantially greater enlargements require a two-lens system (a compound microscope) and the turning-point came in fact with the invention of that instrument. The first two-lens optical systems were the telescopes, and the idea of a compound microscope came essentially from them. In 1610 Galileo made one of the first compound microscopes with the two lenses of a telescope, and in 1611 Kepler worked out the first rules of the new instrument.

The invention of the microscope brought about an immense revolution in science. It led to the discovery of an entirely new world of living creatures that are invisible to the naked eye, the so-called *micro-organisms*. The microscopists of the seventeenth century were the first men who saw bacteria, protozoa, blood cells, spermatozoa and a thousand other *animalcula*, and gradually realised that the large creatures of the visible world are actually a minority in nature. The micro-organisms make up the true major continent of life, and their discovery changed our perception of nature to the very core.

Unfortunately, the microscopes of the seventeenth and eighteenth centuries had a basic structural defect. Lenses that are made of a single piece of glass cannot focus in one point all the light rays that cross them, and their images are inevitably affected by aberrations. The rays that traverse the periphery of the lens, for example, do not converge with those that cross the central part, thus producing a *spherical aberration*. Likewise, the rays which have different colours (or frequencies) converge at different distances from the lens giving origin to *chromatic aberrations*. Because of these distortions, people could see only isolated cells, such as bacteria and protozoa, or plant cells, which are separated by thick cellulose walls, but could not see cells in animal tissues. It is true therefore that in those centuries people saw many types of cells, but the microscope was showing that the smallest units of plants (the compartments that in 1665 Robert Hooke called *"cells"*) are not seen in animals, and it was impossible therefore to think of a common structure.

The discovery that cells exist in all organisms required a new type of microscope, and this came only in the nineteenth century, when the aberration obstacle was overcome by the introduction of *achromatic lenses*. These are made of two or more pieces whose geometrical forms and refraction indices are such that the aberrations of one piece are precisely compensated by those of the other. The first achromatic microscope was build by Giovanni Battista Amici in 1810, and with this new instrument came a systematic revision of all that the microscope had revealed in previous centuries. In 1831 Robert Brown discovered that plant cells contain a roundish refracting mass that he called the *nucleus*, and inside the nucleus it was often possible to see an even more refracting structure that later became known as the *nucleolus*. In 1839 Matthias Schleiden and Theodor Schwann compared plant embryos (which do not have the thick cellulose walls of adult tissues) with animal embryos, and discovered that their microscopic structures are strikingly alike. They are both made of nucleated cells, hence the conclusion that the cell is a universal unit of the living world. This idea brought down the century-old barrier between plants and animals and represents the first part of the cell theory: *all living creatures are made of cells and of cell products.*

Any new idea, however, raises new problems, and in this case the main issue was about the mechanism by which cells are generated, a topic where Schleiden and Schwann made a proposal that turned out to be completely wrong. They suggested that cells originate with a mechanism which is somewhat similar to crystal growth, and which they called *free formation.* The daughter cells were supposed to come from germs or seeds in the nucleus that would grow inside the mother cell like crystals in a saturated solution. The discovery of the true mechanism required many other years of research, and came essentially from embryology studies. In the earliest stages of development it is often possible to see all the cells of an embryo, and, as their number grows, one can realise that they always contain nuclei whose size and shape are practically constant. This means that cells never go through a germ-like stage, where they would have to be smaller than nuclei, and must be produced by a process that keeps their basic structure invariant, i.e. by a process of *replication.*

In 1852 Robert Remak explicitly rejected the free-formation idea and concluded that *"Cells always come from the division of other cells."* In 1855 Rudolf Virchow reached the same conclusion by studying a great number of normal and pathological adult tissues, and condensed it with the motto *"omnis cellula e cellula".* The final version of the cell theory is therefore the combination of Schleiden and Schwann's first theory with the conclusion of Remak and Virchow: *"All living creatures are made of cells and of cell products, and cells are always generated by the division of other cells."*

The problem of generation

At the very centre of biology there are two complementary problems: *"How does an organism produce an egg?"* (the problem of generation), and *"How does an egg produce an organism?"* (the problem of embryonic development). These questions have been debated since antiquity – both Hippocrates and Aristotle wrote at length about them – but only the microscope made it possible, in the nineteenth century, to make the crucial observations that led to a solution.

With the cell theory, organisms became *societies of cells*, and the problem of generation became the problem of understanding which and how many cells are forming the germ of a new individual. Botanists believed that any seed had to be fertilised by a high number of pollen grains, and it was widely held that the greater that number the stronger would be the resulting plant. The same thing applied to animals, where it was again thought that an egg had to be fertilised by many spermatozoa, each carrying a fraction of the hereditary material, because it was *an experimental fact* that an egg is always surrounded by a multitude of spermatozoa at fertilisation, and it was taken for granted that a single spermatozoon could not possibly carry all the hereditary traits of the body.

It was Oskar Hertwig, in 1875, who solved this problem. By studying sea urchins, animals which are particularly suitable for microscopy studies because of their transparency, he noticed that eggs contain a single nucleus before fertilisation and two nuclei immediately afterwards. He realized that the second nucleus had come from a spermatozoon, and therefore that a *single* spermatozoon can fertilize an egg. Hertwig's discovery was completed in 1879 by Hermann Fol, who managed to inject many spermatozoa into a single egg, and found that in this case development is always abnormal, thus proving that fertilisation can and must be realised by a single spermatozoon.

This however was only a first step. The idea that fertilisation is brought about by the union of one spermatozoon and one oocyte is important, but does not solve the problem of generation. We still need to understand why spermatozoa and oocytes are the only cells that are capable of generating a new individual. What is it that gives them such a power? That makes them so different from all other cells of the body? Once again the answer came from microscopy studies, but new techniques had to be developed first. The decisive innovations were more powerful microscopes (microscopes with a higher resolving power) and the deployment of staining techniques. The dye eosin, for example, gives a pink colour to the cytoplasm while haematoxylin makes the nucleus intensely blue, and a high-resolution microscope reveals that the blue dye of the nucleus is concentrated in discrete bodies that were called *chromosomes* (coloured bodies).

The new technology made it possible to discover that chromosomes undergo spectacular conformational changes and elegant movements (the *chromosomes' dance*) during cell division, a process that Walther Flemming called *mitosis*. But the most significant discovery was the demonstration that the entire chromosome set is divided in two identical parts during mitosis, one for each daughter cell, which strongly suggests that chromosomes are the carriers of hereditary characters. At this point there was only one missing piece in the generation puzzle, and that came with a discovery made by Edouard Van Beneden in 1883. Van Beneden found that in the worm *Ascaris* there are four chromosomes in almost all cells, but only two in their sexual cells (the gametes). And he pointed out that maternal and paternal chromosomes are brought together in the fertilised egg (the zygote) to produce again a cell with a full complement of four chromosomes. Van Beneden, however, published the data without comments, and did not ask why gametes have only half the chromosomes of all other cells.

It was August Weismann, in 1884, who understood the meaning of Van Beneden's discovery, and concluded that sexual cells must undergo a very special division that halves their chromosomal set, so that the union of two gametes at fertilisation could restore the normal (diploid) number. This special division was called *meiosis* in order to distinguish it from normal mitosis, and in 1890 Oskar Hertwig proved the experimental reality of meiosis by describing in detail all its phases. This, then, is what distinguishes the sexual cells from all the others and gives them the power to generate a new individual: *only sexual cells divide by meiosis*.

Weismann gave the name of *somatic cells* to those that divide only by mitosis (and are thus destined to die with the body), and called *germinal cells* those that can divide both by mitosis and meiosis. These are potentially immortal, because they can have descendants for an indefinite number of generations.

The discoveries of fertilisation, meiosis and germinal cells, in conclusion, made it possible to give a precise answer to the generation problem in cellular terms: *the generation of a new individual starts with two meioses, when gametes are formed, and is realised at fertilisation, when a zygote is formed.*

The problem of embryonic development

The most elegant experiment in the history of embryology was performed some 2400 years ago by Aristotle. He opened the shell of chicken eggs at different incubation days, and carefully described what he saw: the white spot on the yolk that marks, at the very beginning, the point where the future embryo is going to appear; the tiny brown lump that starts pulsating at the third day and later will turn into a heart; the greatly expanded vesicles that will become eyes; the entangled red vessels that descend into the yolk and branch out like roots; and the thin membrane that wraps everything up like a mantle.

On the basis of these observations, Aristotle concluded that in a developing embryo organs not only increase in size, as Hippocrates had said, but also in number. Embryonic development, according to Aristotle, is an *epigenesis*, a chain of one genesis after another, where new structures and new functions appear at various steps. During embryonic development, in short, *the complexity of the system increases.*

Almost 2000 years later, around 1660, Marcello Malpighi repeated Aristotle's experiment, but with an important difference. He was the first man to watch a developing embryo under a microscope, and what he saw led him to a very different conclusion. The area where blood vessels are destined to appear, for example, is apparently empty to the naked eye, but under the microscope is full of capillaries. Aristotle had concluded that blood vessels appear *ex novo*, but according to Malpighi he had been betrayed by his own eyes. Could he have used a microscope, he would have realised that organic structures are present even when they are not yet visible. Malpighi therefore reached the conclusion that an embryo's development is not an epigenesis but a *preformation*, a growth of forms that already pre-exist in the fertilised egg.

The theory of preformation was enthusiastically accepted by almost all naturalists of the seventeenth and eighteenth centuries. Swammerdam, Leeuwenhoek, Leibnitz, Réaumur, Spallanzani, Boerhaave, von Haller, Bonnet and many other great scientists declared themselves convinced preformationists, and this not only for experimental reasons but mainly for theoretical ones. They did know the laws of geometrical optics and were aware that their microscopes were affected by aberrations, but the existence of living creatures that are invisible to the naked eye could not be disputed, and was leading to an extraordinary conclusion. The great idea of preformationism was the principle that *the infinitely small is as real as the infinitely large*, and this meant that it is always possible to explain living structures with smaller structures. Such a conclusion was indeed legitimate at the time, because there was no atomic theory in physics and chemistry, but once again it was the microscope that decided its destiny.

The technological evolution of microscopy eventually made it possible to observe even the earlier stages of development, and it became clear that very young embryonic structures are totally different from adult ones. In 1828, Karl Ernst von Baer published *On the Development of Animals*, a monumental treatise of comparative embryology that ended once and for all any version of preformationism. Von Baer showed that in animal species there is a common stage of development where the entire embryo is nothing but a few sheets of organic matter, or *germinal layers* (ectoderm, mesoderm and endoderm). And the evolution of these basic structures clearly showed that embryonic development is not only a growth process, but also a continuous emergence of new tissues, and a series of three-dimensional movements that deeply transform the shape of the developing embryo.

With the advent of the cell theory, embryonic growth was immediately accounted for by a sequence of cell divisions. A fertilised egg becomes 2 cells, and then 4, 8, 16, 32, 64 and so on. With 10 divisions the cell number is about a thousand, with 20 is a million, with 30 is a billion, with 40 is a thousand billion, and so forth. For the fifty thousand billion cells of an adult human body, therefore, all that is required is 45-46 cell divisions. The difference between an adult body and a fertilised egg, however, is by no means a mere question of cell numbers. Fifty thousand billion eggs, whatever their arrangement in space, would never make a human being, and it is clear therefore that during development cells must become *different* from the fertilised egg. Embryonic develoment is accompanied therefore by a hierarchy of differentiation processes (which in man produce more than 200 types of cells).

During development, furthermore, the external shape and the internal anatomy of an embryo undergo many transformations before one can start recognising the familiar features of adult life. These changes are brought about by migrations, tubulations, invaginations and foldings of many types, and are collectively known as *morphogenesis.*

The discoveries of cell growth, histological differentiation and morphogenesis, in short, gave a precise answer to the problem of embryonic development in cellular terms. *Embryonic development is a true epigenesis and consists of three fundamental processes: growth, differentiation and morphogenesis.*

The two versions of the cell theory

The great philosophers of antiquity discussed quite a number of world views, such as the atomic theory, determinism and indeterminism, relativity and evolution, and yet none of them conceived the cell theory, which makes us wonder why. The fact that they did not have the microscope does not seem to be decisive from a conceptual point of view. Even atoms cannot be seen, and yet the atomic theory was explicitly formulated. The problem is therefore the following: *Why could ancient people think about atoms but not about cells?* The idea that matter can be divided into particles is suggested by many facts of daily life: a house is made of bricks, a desert is made of grains of sand, drops of rain can be turned into a river, and so on. Why not add that organisms are made of micro-organisms?

The reason is that in this case the ancients found themselves up against an overwhelming obstacle, because experience shows that a mother is always bigger that the embryo which is born from her. Life on Earth must come therefore from above, not from below, from a superior Being – God or Mother Nature – not from small insignificant microbes. In such a situation the microscope was absolutely indispensable to force us to see the cells, to impose their existence on us, because without this violence our minds would never have been able to believe them. The cell theory has undoubtedly been one of the great revolutions in the history of thought, perhaps the greatest of all, and yet one can still hear the suggestion that it is not a real scientific theory because it has a purely descriptive nature. According to this view, the theory is but a record of the empirical fact that all living beings are made of cells, and that every cell derives from a pre-existing one.

In reality, this happens because the cell theory can be expressed either in a *weak* or in a *strong* version. The theory can indeed be reduced to a mere description of life when it is formulated by saying that "*All* known *living organisms are made of cells.*" In this case it has no predictive power and no falsifiable consequence. But there is also a strong version that does represent a true falsifiable generalization of the empirical facts, and therefore a true scientific theory. It is the statement that "*All* possible *living organisms are made of cells.*"

The first version is a mere acknowledgment that cells exist, at least on our planet. The second one states that cells are the fundamental components of *all* forms of life, including extraterrestrial and artificial life. It states that cells are the *logical* units of the living world, just as atoms are the units of the physical world. The strong version of the cell theory, in other words, declares that life does not exist without cells, and represents therefore a definition of life itself: *life is the state of activity of cells and of cellular systems.*

The very first problem of biology, the question *"What is life?"*, becomes therefore *"What is the cell?"* In order to answer this, however, we must recall the answers that have been given in the past to the question *"What is a living organism?"*

Mechanism

There are at least two good reasons for saying that modern biology was born in Europe in the first half of the seventeenth century. One is the discovery of the new world of micro-organisms. The other is the formulation of the first great paradigm of biology: the idea that *every living organism is a machine.* This concept – known as *mechanism* – found in René Descartes its most outstanding advocate, but in reality it was the result of a collective convergence of ideas by scholars of many European cities. From antiquity up to the end of the Renaissance, machines had been built with the sole purpose of obtaining practical benefits, but in the seventeenth century this view was enlarged by two fundamental novelties.

The first is that machines started to be seen not only as a means for changing the world, but also as an instrument for studying it. In order to look into a microscope, and accept the reality of micro-organisms, one must first of all *believe* in what one is seeing, trust that the instrument is not producing optical illusions (as some were saying) but is revealing structures that do exist in the real world. The second novelty of the seventeenth century is that machines became not only an instrument of knowledge but also *a model* of knowledge. The idea was developed that to understand the human body it is necessary to divide it into parts, and to study the functioning of its smaller components, just as we do with machines.

"A healthy man is like a well functioning clock, and an ill man is like a clock that needs repairing." This statement by Descartes is a perfect summary of mechanism, and inspired a radical transformation of medicine. Anatomy ceased to have a purely descriptive role and moved towards physiology and pathology. A physician did not have to rely on the books of Hippocrates and Galen but on experience, as any good mechanic does. The revolution of mechanism cut deeply into every aspect of European thought. Even the concept of God changed, and the Omnipotent became the Supreme Mechanic, the creator of the laws that govern the "machine" of the universe. And God is to universe what man is to machine. This idea inspired a complete separation of thought and matter and found its highest

expression in Descartes' dualism, in the distinction of *res cogitans* and *res extensa*, i.e. in a total divorce of mind from body. It was the beginning of modern philosophy.

It has been said (and it is likely) that no great cultural revolution can be a sudden event. It must necessarily be preceded by a long period of incubation, possibly in unlikely places and by the hands of unimpeachable players. In our case, many historians suspect that the cultural mutation of mechanism appeared in the first centuries of the Christian era, and was nursed in monasteries. In a world that was increasingly falling apart, those were the only places where a remnant of civilisation was kept alive, by cutting bridges with the outside and by living in self-sufficient communities. But in those places economic independence was only a means to the goal of spiritual life, and machines started to be built so that time could be subtracted from labour and dedicated to prayer and meditation.

Machines were no longer instruments of slavery but tools of liberation, a gift from God, and it became important therefore to understand them, to improve them, and to build new ones. The machine culture was particularly nursed in Benedictine abbeys, but gradually it went outside their walls, spread into neighbouring urban communities, and entered the shops of artisans and artists. And finally it also knocked at the doors of universities.

Whatever did happen in those centuries, it is a fact that with the beginning of the seventeenth century a completely new type of machine started appearing: tools that served no practical purpose, and were used only for the demonstration of theoretical principles (a typical example was the inclined plane that Galileo built in order to illustrate the laws of motion). In the eighteenth century, furthermore, machines appeared that were even more useless and bizarre, like Jacques de Vaucanson's mechanical duck (that flapped its wings, quacked, ate and expelled artificial faeces) or the "Writer" of Pierre Jacquet-Droz, an automaton that could dip his pen into an inkwell, shake off the ink and write Descartes' phrase *"Cogito ergo sum"* (Figure 1.1).

These machines were apparently built for amusement, and could easily be mistaken for toys, but in reality they were the equivalent of our artificial intelligence computers. Machines that were announcing the new philosophy with the disarming tools of utopia.

The chemical machine

The mechanical concept of nature spread very quickly in seventeenth century Europe, but not without conflict. Opposition came from virtually all quarters, and it was violent. Apart from the rejection by Aristotelian academics, there was a new science that was slowly

Figure 1.1   The "Writer", built in the middle of the eighteenth century by the Swiss inventor Pierre Jacquet-Droz, is a beautiful automaton sitting at a writing desk that dips his pen into the inkwell, shakes off the excess ink, and writes Descartes' famous motto *"Cogito ergo sum."* The automaton is still fully operational and survives in a Neuchâtel museum.

emerging from the night of alchemy and regarded the human body essentially as a seat of chemical reactions. The heirs of the alchemists were determined to leave magic behind, but had no intention of accepting the "mechanical" view of nature, and one of chemistry's founding fathers, Georg Ernst Stahl (1659-1731), launched an open challenge to mechanism. His thesis was that organisms cannot be machines because they possess a *vis vitalis* that does not exist in the mineral world. Stahl was the first to make a clear distinction between organic and inorganic chemistry, and challenged mechanism with three arguments:

(1) It will never be possible to obtain a synthesis of organic compounds in the laboratory because inorganic materials are devoid of *vis vitalis*.

(2) What is taking place inside living organisms are real transmutations of substances and not movements of wheels, belts and pulleys.

(3) Living organisms cannot be machines because machines do not suffer.

The first objection encouraged a long series of experiments on the *in vitro* synthesis of organic compounds, and was clamorously falsified in 1828, when Friedrich Woehler obtained the synthesis of urea in the laboratory. It is interesting to notice that Woehler himself was a convinced vitalist, and wrote with dismay that he was witnessing *"The great tragedy of science, the slaying of a beautiful hypothesis by an ugly fact"* (this shows that the first vitalists – quite differently from their later followers – fully accepted the principle of experimental falsification).

The second objection of Stahl had a stronger basis, and forced mechanists to change the very definition of *living machine*. In the course of the eighteenth century, in fact, the view that organisms are *mechanical machines* gradually turned into the idea that they are *chemical machines*. This smooth change of perspective went hand in hand with the development of a new engine, an apparatus that was exploiting the chemical reactions of combustion to produce mechanical movements. It was the steam engine that brought together the two sciences, and both mechanists and vitalists realised that a chemical machine is not a contradiction in terms, as had been thought, but a reality.

The third objection of Stahl, the idea that machines do not suffer, has never been overcome, and even today is a major obstacle on the road towards artificial life. Descartes wrote that only human beings suffer because only they have a soul, while animals are merely *mimicking* the expressions of pain, but very few took seriously such an extravagance. It became increasingly clear therefore that an organism cannot be a mere mechanical machine, and eventually the concept of the chemical machine was universally accepted.

In the nineteenth century, the study of the steam engine was pushed all the way up to the highest level of theoretical formalism, and culminated with the discovery of the first two laws of thermodynamics: the principle that energy is neither created or destroyed, and the principle that the disorder (or *entropy*) of any closed system is always on the increase. This second principle had a particularly traumatic impact, because it appeared to expose an irreducible difference between physics and biology. In any closed physical system disorder is always increasing, while living organisms not only preserve but often increase their internal order.

The standard reply that organisms are not closed but open systems is of little comfort, because one needs to understand *how* they manage to keep their highly organized state. Eventually however the answer was found and came from two hypotheses:
(1) Living organisms must continuously exchange matter and energy with the environment (the idea of *biological perpetual motion*).
(2) The internal order of organisms is preserved because the disorder produced by their chemical reactions is continuously pumped outside them.

In order to remain alive, in other words, organisms must be in a perpetual state of activity (their cells work even when they sleep), and must continuously pump out the excess entropy of their reactions. In the words of Erwin Schrödinger (1944), they eat not only matter but also order. Towards the end of the nineteenth century, in conclusion, a living organism came to be seen essentially as a *thermodynamic machine*, i.e. as a chemical machine that must be continuously active in order to obey the laws of thermodynamics.

The computer model

Towards the end of the eighteenth century, just as the chemists' critique was giving way, another opposition to mechanism arose and gave origin to a new version of vitalism. This movement started as a spontaneous, almost instinctive, reaction of many biologists to a veritable absurdity that mechanists wanted to impose on biology. It was a revolt against preformationism, the idea that adult structures are already preformed in a *homunculus* within the fertilised egg. In 1764, Charles Bonnet explicitly launched the great challenge of preformationism: *"If organised bodies are not 'preformed', then they must be 'formed' every day, in virtue of the laws of a special mechanics. Now, I beg you to tell me what mechanics will preside over the formation of a brain, a heart, a lung, and so many other organs?"*

The challenge was clear, and in order to avoid preformationism biologists were forced to conclude that the *formative force* required by Bonnet in order to account for embryonic development must indeed exist. It was an embryological, rather than a chemical, force, very close to Aristotle's *inner project*, but it also was given the name of *vis vitalis*. Preformationism, as we have seen, was definitely abandoned in 1828, when von Baer's monumental treatise showed that embryonic development is a true epigenesis, as Aristotle had maintained, i.e. a genesis of new structures and not a simple growth of pre-existing structures. Once again, mechanists were forced into admitting that the concept of a "living machine" had to be modified in order to account for the reality of embryonic development, but this time a solution turned out much more difficult to find, and throughout the whole of the nineteenth century the claim of vitalism appeared unsurmountable.

The answer came only with genetics, and more precisely with the discovery that life does not consist only of matter and energy but also of *information*. In 1909, Wilhelm Johannsen made a sharp distinction between the visible part of any organism (the *phenotype*) and the part that is carrying hereditary instructions (the *genotype*), and argued that a living being is not a monad but a dual system, a diarchy, a creature that results from the integration of two complementary realities. Unfortunately, Johannsen's message was either ignored or misunderstood, and it was only the computer, with the distinction of *hardware* and *software*, that turned the *phenotype–genotype* duality into a comprehensible and popular concept.

What matters is that the genotype – the biological software – is a deposit of instructions and therefore is potentially capable of carrying the project of embryonic development. This was the long-awaited answer to vitalism, and the computer became therefore the new model of mechanism. In reality, the new model of a living machine is not the computer that we encounter in our daily life, but an ideal machine known as *von Neumann's self-replicating automaton*.

John von Neumann, one of the founding fathers of computer science (it was he who invented the central processing unit), asked himself if it is possible to design an automaton that is capable of building any other automaton (a *universal constructor*), and in particular an automaton that builds copies of itself (a *self-replicating machine*). His great contribution was the demonstration that such machines are *theoretically* possible (Figure 1.2). In practice, a von Neumann's self-replicating machine has never been built because of its complexity (it requires more than 200 000 components), but the proof that it *could* be built amounts to saying that it is possible, and proves therefore that a machine is capable of replication (Marchal, 1998). Von Neumann announced these conclusions in 1948, and his work inspired a completely new research field that today is already divided into disciplines and is collectively known as *artificial life*. A parallel, but different, field is that of *artificial intelligence*, and it is important to keep them apart. Artificial intelligence studies characteristics that in real life appeared at the end of evolution, whereas artificial life simulates what appeared at the beginning (Sipper, 1998; Tempesti *et al.*, 1998).

In the field of artificial life we are today at a level that organic life reached about 4 billion years ago, at the time of the so-called primordial soup, but the interesting thing is that we could actually witness the origin of this new form of life with our own eyes. This is the last frontier of mechanism, the borderline beyond which the dream could become true.

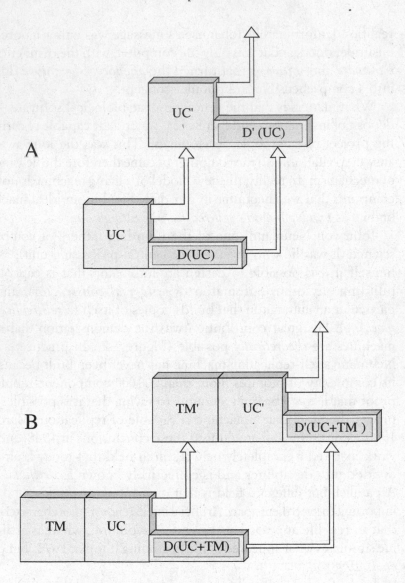

Figure 1.2   Von Neumann's self-replicating machine.
(A) A universal constructor UC can use its own description D(UC) to build a copy of itself, UC', and of its description D'(UC).
(B) A universal constructor UC can include a universal computer, for example a Turing machine (TM), and build a copy of the entire system from its description D(UC+TM).

The autopoietic cell

Artificial life is an entirely new approach to the fundamental problems of biology, because it allows us to study life in a totally different way, i.e. by building machines that have some of its properties. It must be underlined, however, that silicon-based life is utterly different from carbon-based life because artificial molecules and artificial cells are made of electronic circuits and are therefore two-dimensional creatures. This explains why biologists have not abandoned more traditional approaches, and the search for a proper definition of organic life has never stopped. In this field, an important step forward was made in 1974 by Francisco Varela, Humberto Maturana and Ricardo Uribe, with the paper that introduced in biology the concept of *autopoiesis*.

In order to illustrate their idea, Varela and Maturana used the tale (already exploited by Alexander Oparin) of a green man from Mars who comes to Earth and wants to discover what kind of life exists on our planet. He makes a long list of terrestrial objects but is not so sure about their living status, and asks a farmer to help him. The farmer takes a look at the list and immediately divides the objects in two columns, living at the left and not-living at the right:

man	*radio*
tree	*motor car*
mushroom	*computer*
mule	*robot*
hen	*moon*
coral	*tide*

The green man is surprised by such a display of confidence, and asks the farmer to tell him by which feature he could pick up the living so quickly. The farmer takes two objects at random – mule and hen – and say that they are alive because are capable of *"movement"*, but the green man is not convinced. Coral and tree do not move but are definitely alive. At that point the farmer suggests *"irritability"*, or *"the ability to react to stimuli"*, but again the answer fails. It is true that man and mule react to a needle's puncture, but tree and coral remain indifferent. The farmer then shouts *"Reproduction!"* but immediately has to change his mind because the mule does not reproduce. And yet his two columns are absolutely correct. But why? What is it that he knows without being aware of knowing? The farmer needs to think it over, and asks the green man to come back the next day. Then he starts thinking.

Trees lose their leaves in autumn, and produce them again in springtime, by growing new ones from the inside. And animal hair does the same thing: it grows from within. The farmer knows that when he is starving his body weakens and becomes thinner, but as soon as he starts eating again his growth goes back to normal. And it is always an internal activity that keeps the body growing. At this point he has understood, and is ready to answer the green man.

All the objects of the right column – radio, motor car, etc. – are not capable of repairing themselves, while those of the left column are alive precisely because they have this property. Now the green man is satisfied and agrees with him. The second principle of thermodynamics had been discovered even on Mars, and green men knew that an organism must be in a perpetual state of activity in order to be alive. Not only must a body be capable of repairing itself when something breaks down, but it must be repairing itself all the time, it must always be demolishing and rebuilding its own structures, i.e. it must be capable of *permanent self-production*, or *autopoiesis*.

Varela and Maturana add that autopoiesis must be a property of every living system, including its smallest units, which means that any cell can be represented by a scheme that illustrates its continuous transformation of external matter into cellular components (Figure 1.3). When production is equal to demolition, a cell is in a stationary state (*self-maintenance*); when production is greater than demolition, a cell grows and eventually divides itself into two (*self-reproduction*). Varela and Maturana arrive in this way at a definition of the living system in general and of the cell in particular: *a physical system is alive if it is capable of transforming a flux of external matter and energy into an internal flux of self-maintenance and self-reproduction.*

This definition, as we have seen, is an automatic consequence of

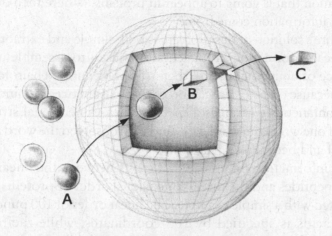

Figure 1.3   The autopoietic cell. In this simplified scheme, molecules from the outside environment (A) are transformed into cellular components (B), while degraded compounds (C) are expelled. When the production rate equals the degradation rate, a cell is in a stationary state. When production exceeds decay, a cell grows and then divides in two.

the second principle of thermodynamics, and the *autopoietic machine* represents therefore an updated version of the nineteenth century concept of the *thermodynamic machine*. But is this really the most general definition of life?

The epigenetic cell

One of the greatest biological achievements of the twentieth century was the discovery that the information of a gene is determined by the order of its nucleotides, pretty much as the information of a word is due to the order of its letters. In both cases information corresponds to the order of elementary units along a line. Genetic information is therefore a *linear* quantity, but the function of proteins is determined by the arrangement of their amino acids in space, i.e. by their *three-dimensional* information. Clearly genes are not transporting all the information that is going to appear in proteins. Where then does the missing information come from?

Nature's solution of this problem is both simple and extraordinary. The linear information of nucleotides is used to assemble a linear sequence of amino acids, and then this polypeptide chain folds on itself (because of the electrical forces that exist between amino acids) and spontaneously assumes a specific three-dimensional structure. It is as if one wrote the word *apple* and then observed the word folding on itself and becoming a real apple.

The information difference that exists between the linear order of polypeptides and the three-dimensional order of proteins can be illustrated with a simple example. The linear order of 100 punctiform amino acids is specified by 100 coordinates, while their three-dimensional organisation requires 300 coordinates (three for each amino acid). Protein folding, or self-assembly, amounts therefore to adding the 200 missing coordinates to the 100 coordinates provided by the genes. And since the complexity of a system is determined by the number of parameters that are required to describe it, it is clear that protein folding is a phenomenon that produces *an increase of complexity.*

In embryonic development, as we have seen, the term *epigenesis* has been used to describe the increase of complexity that takes place in a growing embryo, but that term can be generalised to any other convergent increase of complexity, and we can say therefore that protein folding is an example of *molecular epigenesis.* The three-dimensional information of a protein "emerges" during folding exactly as the properties of water "emerge" from suitable combinations of hydrogen and oxygen atoms.

But the folding of linear polypeptides into three-dimensional proteins is only the first of a long series of assemblies. Once formed, some proteins are assembled into larger aggregates, and these in turn give rise to higher structures. The enormous amount of information which is stored in the three-dimensional structure of a cell comes therefore from a chain of assemblies, and all these processes are *epigenetic*, not only because they take place after the expression of genes, but also because new properties emerge in stages very much like the way in which novelties appear in embryonic development. The structure of a cell, in conclusion, is the result of a transcription of genes and of a series of different types of assembly that collectively represent a trūe *cellular epigenesis*. Genetic information is invariably at the beginning of all these steps, but the three-dimensional information of a cell is vastly greater, and is almost entirely due to the increase of complexity produced by cellular epigenesis. As strange as it may be, however, among the properties that have been proposed to define a living system (heredity, metabolism, reproduction, homeostasis, adaptability, autopoiesis, etc.), *epigenesis has never been mentioned* (see Appendix).

Perhaps the explanation is that in theory (but only in theory) we could do without it. If a genotype contained a *complete* description of the phenotype, there would be no need for an increase of complexity, and no need for cellular epigenesis, as is the case in all our machines, from mechanical clocks to Von Neumann's automata. But nature has adopted a totally different strategy. Even in the most simple bacterium, the genome does not contain a complete description of the cell, but only the linear information of its polypeptides, each of which acquires a three-dimensional form with an assembly process that is not written in the genes. This is the great difference between cell and machine. No man-made machine is capable of increasing its own complexity, and it is precisely cellular epigenesis that makes a living cell qualitatively different from any known machine.

A proper definition of the cell cannot ignore this fundamental characteristic, and must mention it *explicitly*. We arrive in this way at a new definition of the cell that can be expressed in various ways:

(1) The cell is an autopoietic system whose thrcc-dimensional structures are built by assemblies that increase its complexity.

(2) The cell is an autopoietic system where the phenotype is more complex that the genotype.

(3) The cell is an autopoietic and epigenetic system.

The increase of complexity, in conclusion, is a qualifying property of life. This requires a new definition of the cell and will allow us, as we will see, to discover a new logic at the basis of life.

2

THEORIES OF EVOLUTION

Our view of evolution has gone through various building stages in the last two centuries, and this chapter presents a bird's-eye view of those steps. It starts by emphasising that for the two founding fathers – Lamarck and Darwin – evolution was necessary not to explain the past but to understand the present. More precisely, it was the only way to account for the experimental fact that organisms are admirably adapted to their niches and lifestyles. In the end, it was Darwin's idea of natural selection that solved the problem, and today this is virtually a closed chapter: adaptation is always the result of natural selection. Adaptation, however, is not everything in life, and the existence of other mechanisms has been repeatedly debated. The best example is genetic drift, whose reality is now beyond doubt and which clearly is a second mechanism of evolution at the molecular level, even if the relative contribution of natural selection and neutral drift at this level remains an open problem. A third mechanism of molecular evolution has been inspired by Barbara McClintock's work and can be referred to as *evolution by genomic flux*, or, in Gabriel Dover's terminology, *by molecular drive*, but in this case the consensus is not yet widespread. Another topical issue in evolutionary biology has been the unification of all life sciences under the principle of natural selection, a project that up until recently was called the *Modern Synthesis*, or *panselectionism*, while today it is generally referred to as *universal Darwinism*. The plain truth, however, is that the Modern Synthesis has never achieved the unification of evolution with embryonic development, and this makes it legitimate to think again about its basic assumption.

Traditional biology

The first biology books were written about 2400 years ago by Hippocrates and by Aristotle, and in those volumes we find not only a detailed account of all that was known at the time, but also a grand attempt to build a comprehensive view of nature. It is still debatable whether those books can be regarded as the starting-point of biology, but it is certainly true that they were the end result of a long oral tradition whose origins are lost in the night of prehistory.

The plants and the animals of agricultural civilisation had been produced by a collective experiment which lasted thousands of years, and all the results obtained by farmers and breeders were leading to a precise general conclusion: *it is possible to produce new varieties of plants and animals, but it has never been possible to produce new species.* This was the meaning of apparently naive statements such as *"Daisies only come from daisies and elephants only from elephants."*

Aristotle's writings on anatomy, physiology and animal behaviour show that he had many contacts with breeders, farmers and fishermen, and perhaps it was their testimony that made him reject the historical transformation of organisms, an idea that was fashionable in his times and that even Plato had accepted (as a continuous degeneration). Eventually, however, the conclusion that *"Species are immutable"* did prevail, and even received a religious blessing, but it would be wrong to forget that its real basis was the millennial experience of farmers.

The weight of this idea comes from its consequences. If species are immutable and the world is not eternal, we are bound to conclude that sometime in the past there must have been something very similar to the Creation described in the Bible. At the most, one could say that the days of Genesis had been geological aeons, but the substance would not change. The concept of immutable species leads inevitably, in a finite universe, to the concept of creation.

In the 1600s and 1700s the naturalists started again that *"dialogue with nature"* which had been interrupted shortly after Aristotle's death, and they too realised that the immutability of species was not only a religious dogma, but, above all, the only legitimate conclusion that one could come to from the evidence of farmers and breeders. It was the generalisation of countless experiments and in this sense it was a true scientific hypothesis.

In the seventeenth and eighteenth centuries, moreover, the invention of the microscope and the first great geographical explorations proved that living creatures were far more numerous and varied than people had thought in the past. In his *Historia Animalium*, Aristotle described nearly 550 animal species, whereas, in 1758, Linnaeus listed more than 4000 and wrote that the actual number must be at least a hundred times greater. The discovery of this enormous *diversity of life*, however, did not raise any conflict with religion. One could be puzzled by the fact that insects were by far the most numerous inhabitants of the Earth, but the Creation idea was not constrained by numbers, and the discovery of new species could always be interpreted as yet another proof of the Creation's magnificence.

In addition to the extraordinary diversity of life, the naturalists of the seventeenth and eighteenth centuries made another fundamental discovery. The plants and the animals of different parts of the world could differ in virtually every character, but all had one thing in common: every one was perfectly adapted to its environment. *The woodpecker has feet, tail, beak and tongue which are admirably adapted to capturing insects beneath the tree bark; the webbed feet of ducks and geese are clearly made for swimming; bats go hunting by night and use ultrasonic waves to locate their prey and to avoid obstacles.* Countless examples showed that *adaptation to the environment* was a universal phenomenon, and the naturalists rightly concluded that it must be a general property of life. In this case too the discovery could be seen as a proof of divine omniscience and did not lead to any conflict with religion.

The immutability of species, the diversity of life and adaptation to the environment, in conclusion, are general concepts that were inferred from countless experimental discoveries, and the fact that they were seen as products of a divine project does not in the least diminish their importance. Traditional biology was built almost entirely by profoundly religious men, and we must be grateful to them for the ideas that they left us.

Lamarck's contribution

The great geographical explorations of the seventeenth and eighteenth centuries revealed not only the existence of an entirely new world of plants and animals, but also the proof of immense geological transformations. Rocks that once had been at the bottom of the sea (as the presence of fossil shells showed) could be found on mountain tops. Vast temperate regions carried signs that once they were occupied by glaciers, and territories that volcanoes had covered with lava were now fertile and full of life. The world had clearly gone through a turbulent history, but even this discovery did not lead to any conflict with tradition. Many, in fact, saw in it the proof of the deluge and of the other catastrophes described in the Bible.

The eighteenth century, however, was also the time of the Enlightenment and in the newly found freedom of thought various theories appeared on the *transformation* of species, but those speculations were merely a return to the ideas of Greek philosophers, and rejected a priori the ideas of traditional biology without a comparable experimental basis. Even the great David Hume thought he could demolish one of the pillars of traditional biology (the concept of adaptation) but he was mistaken. His thesis was that adaptation is a false problem because organisms could not live if they were not adapted, which amounts to saying that adaptation does not need an explanation because it is a universal feature of life, while it is precisely because of this that it must be explained.

Today the idea of evolution is so tightly linked to the evidence of the fossil record that we can hardly appreciate how difficult it was to overcome the traditional view without that evidence (which in the eighteenth century was almost nonexistent). And yet the idea of evolution was born precisely in this way: as a theory that we need to explain not the fossils but what we see *today* around us. One cannot insist enough on this point: the theory of evolution was not proposed to explain the past, but *to understand the present*.

The man who proposed it, in 1809, was Jean Baptiste Lamarck, and his great contribution was a simple but shattering idea: the facts of biology and geology are compatible with the traditional view when examined *one by one*, but this is no longer true when they are examined *together*. If the surface of the Earth has undergone the physical changes described by geology, it is no longer possible to say that species are always adapted to their environment and to say at the same time that species do not change when their environments change. One can either renounce the geological changes, or adaptation, or the fixity of species. The three things *cannot be simultaneously true*: this was Lamarck's great intuition.

He concluded that the ideas to be trusted are those for which we have direct and undisputable evidence, i.e. (1) the diversity of life, (2) adaptation to the environment and (3) the geological changes. What must be abandoned is the fixity of species, because its only empirical basis comes from the experiments of farmers and breeders, and these started only a few thousand years ago while life's history, according to Lamarck, was far older than that.

Together with the problem of evolution, however, Lamarck also addressed the problem of its mechanism, and here he made hypotheses that turned out wrong, and for which he has been reproached ever since. Lamarck's mechanism (or, rather, what has become known as such) comprises (1) spontaneous generation, (2) an intrinsic tendency towards complexity, and (3) the inheritance of acquired characters. Today all these concepts are rejected, but the reasoning that brought Lamarck to proclaim the reality of evolution is independent of them and remains perfectly valid.

The diversity of life, the reality of adaptation and the geological changes, together, require a world which is incompatible with the fixity of species, and we are bound to conclude that in the past species must have changed. This is Lamarck's real message: evolution is a reality because it is the only idea that explains the world which we see around us. The mechanism is important to understand *how* evolution happened, but in order to prove that it did happen Lamarck's argument is enough. Darwin himself, as we will see, came to this conclusion.

Darwin's bet

Charles Darwin was born in the same year (1809) that Lamarck published, in *Philosophie Zoologique*, the first complete theory of evolution. Darwin became familiar with the transformation-of-species idea since his youth, because his grandfather Erasmus had written a poem on it, and his father Robert, besides a successful doctor, was a declared unbeliever, but their arguments did not convice him. After an attempt at studying medicine (demanded by his father) the young Darwin decided to follow his desire to became a naturalist, and at 19 he entered the theology faculty at Cambridge.

In those days, the great majority of naturalists were church ministers, and the professors that Darwin met in Cambridge (in particular John Henslow and Adam Sedgwick) reinforced his belief that the religious explanation of nature – or *natural theology* – was a world-view far more rational and scientific than the speculations of the transformists. Shortly after the end of his studies (in 1831) Darwin boarded HMS *Beagle* for a five-year voyage around the world, and there is no doubt that at the beginning of the voyage he was a firm believer, as he wrote in his autobiography, in *"the strict and literal truth of every word in the Bible"*.

At the end of the voyage, instead, Darwin was full of doubts, and a year later, in 1837, became an evolutionist, even if he had not yet discovered the mechanism of natural selection, an idea that came to him after another year of meditations, in 1838. These biographical notes are important because they make us understand better than many other discourses how difficult it was, at the time, to believe in evolution on the basis of Lamarck's arguments. Darwin needed to see with his own eyes the effects of geological transformations, the incredible diversity of life forms and, above all, the specialised adaptations of organisms to widely different environmental conditions. These were the very points of Lamarck, of course, but books had not convinced the young Darwin, whereas the impact with nature was much more traumatic. Darwin discovered that his certainties were very fragile constructions, but had the intellectual honesty to admit it, and towards the end of the voyage around the world he came up with what amounted to a personal bet with nature.

During a visit to the Galápagos, in the autumn of 1835, Darwin collected samples of mockingbirds from four different islands, and noted that each group had slight differences not only from those of the other islands but also from the mockingbirds of South America. The ancestors of those birds had surely arrived from the continent, and in each island had developed individual features, a phenomenon quite similar to what is commonly observed in many varieties of domesticated animals that are raised in different geographical regions. But were the Galápagos mockingbirds only different varieties or different species? In July 1836, Darwin wrote in his shipboard diary: *"I must suspect they are varieties ... but if there is the slightest foundation for these remarks, the zoology of the Archipelago will be well worth examining: for such facts would undermine the stability of species."* The bet was clear: if the four types of mockingbirds were mere varieties, Darwin was still prepared to believe in the fixity of species, but if they turned out to be different species he would have to yield to the evidence.

Shortly after his return to England, which took place in October 1836, Darwin sent many samples of the animals he had collected during the voyage to various specialists of the Zoological Society, and in March 1837 he travelled to London to hear their verdict. One of them, the ornithologist John Gould, informed him that three of the four mockingbirds were definitely different species and not mere varieties. Gould, in addition, told him that twenty-five of his twenty-six terricolous birds from the Galápagos were new species, and that the finches he had collected in the Archipelago belonged to thirteen different species.

It was at that point that Darwin became an evolutionist. As Lamarck before him, he discovered that evolution was needed in order to understand the present, to explain the diversity and the adaptations that we see today in the world around us. And, as in Lamarck's case, the reality of evolution could be grasped even if one did not understand its mechanism. The problem of the mechanism, however, remained, and Darwin started thinking about it.

Natural selection

In his *Autobiography* (1876) Darwin wrote that the idea of natural selection came to him essentially from two sources: from his talks with animal breeders, and from the *Essay on the Principle of Population* by the Reverend Thomas Malthus (1798). Darwin added that he had the idea in October 1838, which has made some scholars wonder why he waited 21 years before publishing it (and would have waited even longer if it had not been for Alfred Wallace). In reality there is no mystery in that delay. Darwin postponed publication because *On the Origin of Species* had to deal with a great many consequences of natural selection for the history of life, and he wanted to argue them at length and to illustrate them with as many experimental facts as possible. Even in the *Origin*, at any rate, Darwin states that natural selection is the inevitable conclusion of four *"undisputable"* facts, two from Malthus and two from the breeders. From Malthus he obtained these two conclusions:

(1) All populations can grow at an exponential rate.

(2) The limited resources of the environment allow only a restricted growth.

The automatic consequence of these two facts is that in any population only some can survive, and we therefore have the problem of understanding what it is that decides their survival. Chance? Destiny? A priori we cannot exclude that surviving is a question of luck, and in isolated cases this can indeed happen. Statistically, however, the explanation does not seem likely, and if it were true it would leave us in the most complete darkness.

It is at this point that Darwin resorts to the two undisputable facts from the breeders:

(3) Every animal is a unique individual, in the sense that it is always possible to recognise some characteristics that distinguish it from the other animals of the group.

(4) Of all individual features that distinguish an animal from the others, many are found again in its descendants, and are therefore inherited. These two facts are precisely what allows breeders to practise the artificial selection of animals with great success, and Darwin concluded

that there can be no doubt about them. But these facts amount to saying that in any population not all individuals have the same potential to overcome life's difficulties, and this automatically solves the survival problem: the subjects whose individual characteristics are more suitable to overcome crises have a greater chance of surviving.

In nature there is an automatic selection which is continuously going on simply because variations exist between individual organisms (the breeders' fact), and because not all can survive in a limited environment (Malthus' fact). Darwin described the principle of natural selection at the end of Chapter 4 of *On the Origin of Species*, and it may be useful to read it in his own words:

"If under changing conditions of life organic beings present individual differences in almost every part of their structure, and this cannot be disputed; if there be, owing to their geometrical rate of increase, a severe struggle for life at some age, season, or year, and this certainly cannot be disputed; then, considering the infinite complexity of the relations of all organic beings to each other and to their conditions of life, causing an infinite diversity in structure, constitution, and habits, to be advantageous to them, it would be a most extraordinary fact if no variations had ever occurred useful to each being's own welfare, in the same manner as so many variations have occurred useful to man. But if variations useful to any organic being ever do occur, assuredly individuals thus characterised will have the best chance of being preserved in the struggle for life; and from the strong principle of inheritance, these will tend to produce offspring similarly characterised. This principle of preservation, or the survival of the fittest, I have called Natural Selection."

Organs of extreme perfection

One of the books which most impressed the young Darwin was the treatise of theologian William Paley, *Natural Theology: Or Evidences of the Existence and Attributes of the Deity, Collected from the Appearances of Nature* (1802). The main point was put forward in this way:

"In crossing a heath, suppose I pitched my foot against a stone, and

*were asked how the stone came to be there; I might possibly answer
that, for anything I knew to the contrary, it had lain there for ever: nor
would it perhaps be very easy to show the absurdity of this answer. But
suppose I had found a watch upon the ground, and it should be inquired
how the watch happened to be in that place; I should hardly think of
the answer which I had before given, that for anything I knew, the
watch might have always been there ... no, the answer is that the watch
must have had a maker: that there must have existed, at some time, and
at some place or other, an artificer or artificers, who formed it for the
purpose which we find it actually to answer; who comprehended its
construction, and designed its use ... Every indication of contrivance,
every manifestation of design, which existed in the watch, exists in the
works of nature; with the difference, on the side of nature, of being
greater or more, and in a degree which exceeds all computation."*

Paley goes on with a discussion of anatomical organs, and in the
case of the eye he finds it perfectly natural comparing it with the
telescope. They are both optical instruments, and Paley concludes
that *"there is precisely the same proof that the eye was made for vision
as there is that the telescope was made for assisting it"*. The organs
have a purpose and therefore must have been designed for it. This is
the point that Paley drives home countless times, and which he
presents as the foundation to the whole edifice of natural theology.
Darwin was greatly impressed by the eye example, and in order to
answer Paley's argument he realised that it is necessary to distinguish
between two different aspects of natural selection.

The selection which is commonly performed by animal breeders
is largely a *negative* one, in the sense that it is the elimination of
individuals which are either disabled or which are less endowed with
a desired feature, and there is no doubt that a similar removal of the
weak can also take place in nature. *Positive selection*, instead, is a
process that goes beyond the elimination of imperfections and *adds*
novelties to existing structures by accumulating chance variations.
Negative selection, in other words, does not produce anything new
while positive selection can do precisely that. Clearly, an organ as
complex as the eye can be a result of evolution only if it was produced
by a process of positive selection, and Darwin asked himself if we do

have elements for believing that such a process can actually have occurred in nature. He addresses this problem in Chapter 6 of the *Origin*, in the paragraph intitled "Organs of extreme perfection and complication", and writes the following:

"The simplest organ which can be called an eye consists of an optic nerve, surrounded by pigment-cells and covered by translucent skin, but without any lens or other refractive body. We may, however, according to M. Jourdain, descend even a step lower and find aggregates of pigment-cells, apparently serving as organs of vision, without any nerves, and resting merely on sarcodic tissue. Eyes of the above simple nature are not capable of distinct vision, and serve only to distinguish light from darkness. In certain star-fishes, small depressions in the layer of pigment which surrounds the nerve are filled, as described by the author just quoted, with transparent gelatinous matter, projecting with a convex surface, like the cornea in the higher animals. He suggests that this serves not to form an image, but only to concentrate the luminous rays and render their perception more easy. In this concentration of the rays we gain the first and by far the most important step towards the formation of a true, picture-forming eye; for we have only to place the naked extremity of the optic nerve, which in some of the lower animals lies deeply buried in the body, and in some near the surface, at the right distance from the concentrating apparatus, and an image will be formed on it.

In the great class of the Articulata, we may start from an optic nerve simply coated with pigment, the latter sometimes forming a sort of pupil, but destitute of a lens or other optical contrivance. With insects it is known that the numerous facets on the cornea of their great compound eye form true lenses, and that the cones include curiously modified nervous filaments. But these organs in the Articulata are so much diversified that Müller formerly made three main classes with seven subdivisions, besides a fourth main class of aggregated simple eyes.

When we reflect on these facts ... the difficulty ceases to be very great in believing that natural selection may have converted the simple apparatus of an optic nerve, coated with pigment and invested by transparent membrane, into an optical instrument as perfect as is possessed by any member of the Articulate Class."

Darwin concludes that natural selection can indeed offer another solution to Paley's problem. An organ of extreme perfection can be built *quickly* from a design, as artisans do, but can also be built *slowly* by a natural process of selection if aeons of time are available. This conclusion, however, is not an entirely satisfactory one, because the final result is the same and the two solutions appear to be equivalent. Darwin therefore asks himself if there are experimental arguments that allow us to choose between Paley's solution of the Divine Artisan and the solution of natural selection. And he finds one of the most convincing points in the innumerable imperfections that exist even in apparently perfect organs.

In designing an eye, for example, no engineer would dream of putting the light detectors on the retina *inside out*, i.e. with their sensitive ends away from the light, and yet this is precisely what we find even in the most sophisticated eyes. Countless examples of this type show that the structure of organs is what could be expected not from the execution of a design but from the patient superposition of successive experiments that never start again from zero, and continue therefore to carry traces of the first attempts. At this point Paley's argument from natural theology, which so much impressed Darwin in his youth, becomes very unconvincing, and our reason is bound to conclude that the humble explanation of natural selection does have the disarming semblance of the truth.

Common descent

The publication of *On the Origin of Species* was an immediate success, and the theory of evolution by natural selection was recognised from the start as one of the greatest triumphs of the human mind. At his death, Darwin was buried in Westminster Abbey, near great thinkers of the past such as Newton and Hume, even though his theory was regarded as a danger to religion.

It has been said that there is a paradox in these honours because Darwin did not invent the idea of evolution nor that of natural selection, as he himself openly states in the "Historical sketch" that he wrote for the 6th edition of *On the Origin of Species*. As for evolution, Darwin admits without hesitation Lamarck's priority: *"Lamarck was the first man whose conclusions on the subject excited much attention. This justly-celebrated naturalist ... upholds the doctrine that species, including man, are descended from other species."* With equal fairness, Darwin adds that the idea of natural selection had already been proposed by William Wells in 1813 and by Patrick Matthew in 1831 (another precursor was Edward Blyth in 1835).

In reality, what deeply impressed people was not the idea of natural selection as such, but the abyssal divide that emerged between the simplicity of the idea and the enormity of its consequences. With an ordinary mechanism Darwin arrived at extraordinary conclusions, and these were so radical that nobody could remain indifferent. There is therefore no paradox in the success of the book and in the honours bestowed on its author.

Darwin did propose, however, some truly original ideas, and perhaps the most extraordinary of all is the concept of *common descent*, the theory that all living creatures of our planet derive from a single stock of primordial forms. In Darwin's times, the fixity of species was still the official theory of biology, and generations of past naturalists had built, within that reference system, a grandiose classification scheme that appeared capable of revealing, as Linnaeus put it, *"the Plan of Creation"*.

Dogs and wolves, for example, are different species because they are reproductively isolated, but they have countless other features in common and for this are classified in the same genus *Canis* (*Canis familiaris* and *Canis lupus*). In the same way, tigers and lions are two species of the genus *Panthera* (*Panthera tigris* and *Panthera leo*), as polar bears and grizzly bears are different species of the genus *Ursus*. Tigers and domestic cats, on the other hand, cannot be put in the same genus, but still have so many characters in common that they are classified, together with lions, in a single family (the Felidae).

Dogs, wolves, tigers, lions, cats and bears, on the other hand, are all characterized by meat-eating structures, and for this are grouped together in the order Carnivora. Animals like bats, monkeys and whales are classified in quite different orders and yet they share with carnivores the ability to feed their young by mammary glands, and all animals that have this property are united in the class Mammalia.

Similar criteria allow us to recognise at least five distinct animal classes – mammals, birds, reptiles, amphibians and fishes – but even in these very different groups it is possible to recognize a common feature. Their embryonic development is organised around a dorsal chord (or chordomesoderm), and this allows us to conclude that the five classes belong to a single phylum (Chordata, subphylum Vertebrata). Vertebrates and invertebrates, in turn, have a number of features that separate them from plants, and for this are grouped together in kingdom Animalia.

As we can see, there are different levels of biological features, and this allows us to recognise seven great categories of living organisms, the so-called *taxa* or *taxonomical groups*: species, genus, family, order, class, phylum and kingdom (Figure 2.1).

This impressive system had been built by generations of naturalists within the classical framework of the fixity of species, and was a good description of the order that we see in nature, but the description could also be interpreted in a different way. Taxonomical relationships, for example, could be a consequence of ancestral relationships, and in some cases the signs of parenthood were quite evident. That dogs and wolves had a common ancestor, for example, was easily acceptable, and also fairly obvious was a relationship between cats and tigers. But to say that butterflies had something in common with whales and coconuts, and that all derived from a common ancestor, is quite a different thing. It is important to notice that this idea is not an automatic consequence of evolution. According to Lamarck, for example, the diversity of life was caused by the spontaneous generation of countless lines of descent which arose independently and were not linked by heredity. Common descent was too improbable to be taken into consideration, and no one before Darwin had proposed it. Darwin described it briefly in the last chapter of *On the Origin of Species* with these words:

"I believe that animals have descended from at most only four or five progenitors, and plants from an equal or lesser number. Analogy would lead me one step further, namely, to the belief that all animals and

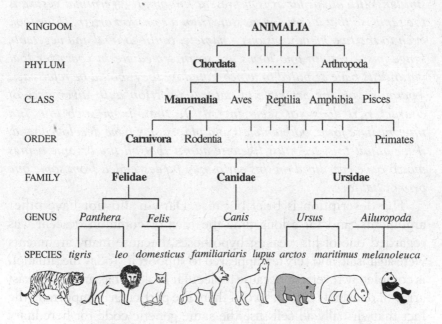

KINGDOM					**ANIMALIA**				
PHYLUM				**Chordata**	·········	Arthropoda			
CLASS			**Mammalia**	Aves	Reptilia	Amphibia	Pisces		
ORDER		**Carnivora**	Rodentia	·····················		Primates			
FAMILY		**Felidae**		**Canidae**		**Ursidae**			
GENUS	*Panthera*	*Felis*		*Canis*		*Ursus*	*Ailuropoda*		
SPECIES	*tigris*	*leo* *domesticus*	*familiariaris* *lupus*	*arctos*	*maritimus*	*melanoleuca*			

Figure 2.1   Organisms are classified into seven taxonomic groups, or *taxa*, which are known as species, genus, family, order, class, phylum and kingdom. Modern taxonomy is still using the binomial terminology introduced by Linnaeus (1758), in which every living form is identified by two Latin names written in italics, the first for the genus and the second for the species. A further subdivision of the seven basic groups is obtained by introducing intermediate categories such as subclasses or superfamilies.

plants have descended from some one prototype. But analogy may be a deceitful guide. Nevertheless all living things have much in common, in their chemical composition, their cellular structure, their laws of growth and their liability to injurious influences. We see this even in so trifling a fact as that the same poison often similarly affects plants and animals; or that the poison secreted by the gall-fly produces monstrous growths on the wild rose or oak-tree. With all organic beings, excepting perhaps some of the very lowest, sexual reproduction seems to be essentially similar. With all, as far as is at present known, the germinal vesicle is the same; so that all organisms start from a common origin. If we look even to the two main divisions – namely, to the animal and vegetable kingdoms – certain low forms are so intermediate in character that naturalists have disputed to which kingdom they should be referred... Therefore, on the principle of natural selection with divergence of character, it does not seem incredible that, from such low and intermediate form, both animals and plants may have developed; and, if we admit this, we must likewise admit that all the organic beings which have ever lived on this earth may be descended from some one primordial form."

The description is brief because Darwin did not have other arguments, and for a long time the idea of common descent was regarded one of his weakest hypotheses, because many arguments were against it, and very few appeared to support it. Today the situation is completely reversed, because molecular biology has amassed a vast array of data in its favour. Two of them are particularly important: the fact that virtually all cells use the same genetic code for hereditary information, and the same molecular carrier (ATP) for energy exchanges. The existence of the same mechanisms in processes as diverse as heredity and metabolism, which are the very foundations of life, can only be explained with the parenthood of all present creatures with all past living beings.

This is probably the greatest of Darwin's ideas, and almost a century later, in 1949, one of the founding fathers of bioethics, Aldo Leopold, underlined its enormous value with this comment:

"It is a century now since Darwin gave us the first glimpse of the origin of species. We know now what was unknown to all the preceding caravan of generations: that men are only fellow-voyagers with other creatures in the odyssey of evolution. This new knowledge should have given us, by this time, a sense of kinship with fellow creatures; a wish to live and let live; a sense of wonder over the magnitude and duration of the biotic enterprise."

The second mechanism of evolution

In Darwin's times heredity was a mystery, but this did not prevent him from concluding that natural selection works on heritable variations. All that he needed to know about heredity were the two facts that he learned from breeders, namely that (1) every individual in a population has unique characteristics, and (2) many distinctive traits are inherited. The discovery of the hereditary mechanism could not cancel these experimental facts, and could not therefore deny natural selection. That discovery, however, could reveal new mechanisms of evolution, and reduce the role that natural selection played in the history of life. This is why the study of heredity came to be seen as the testing ground for any evolutionary theory, and for almost a century, in fact, the debate on evolution has largely been a debate on genetics.

Modern genetics began in 1900 with the rediscovery of Mendel's laws and with the demonstration that hereditary characters behave as discrete instructions carried by material bodies (what Wilhelm Johannsen in 1909 called *genes*). In order to explain the crossing results obtained in the garden of his monastery, Gregor Mendel proposed that every hereditary character (every gene) is determined by two factors (alleles), and that such factors are first separated and then recombined at random in both male and female gametes. On top of that, Mendel proposed that the two alleles of each character can be different (*A* and *a*) and that the final results of their combinations can be of two types only, namely dominant (*AA*, *Aa* and *aA*) or recessive (*aa*).

In 1902, Walter Sutton and Theodor Boveri were able to show that the Mendelian characters (the genes) are physically carried by chromosomes (the *chromosome theory of heredity*), and the study of meiotic divisions and gametogenesis proved that Mendel's hypotheses were absolutely correct, so much so that they could be regarded no longer as hypotheses but as experimental realities. Mendel's laws gave a direct support to the conclusions that Darwin had obtained from the breeders. Every individual is indeed unique, because the recombination of its genes is a totally random process.

The constant reshuffling of recombination, furthermore, means that the genetic variability of any population is virtually unlimited and continuously renewable.

This appeared to suggest that recombination, or *crossing-over*, could be sufficient on its own to change the genetic pool of a population, and therefore to produce evolution, because it seemed that dominant genes would inevitably tend to replace recessive ones. In 1908, however, Godfrey Hardy and Wilhelm Weinberg separately proved that this is an impossible outcome, in ideal conditions, because of a theorem that was soon to become the basis of population genetics. The theorem proved that *"If a population is very large and if no disturbing forces or factors exist, recombination does not change the relative frequencies of the alleles and the genetic pool of the population remains constant."*

The Hardy–Weinberg law is, in a sense, the mathematical equivalence of the fixity of species, and its great merit was to convince geneticists that an evolutionary change can take place only if at least one of its two premises is not fulfilled, that is to say (1) if the population is small, or (2) if disturbing forces or factors are present. The forces or factors which can perturb a genetic pool are the two elements outlined by Darwin – mutations and natural selection – but the Hardy–Weinberg theorem predicts evolutionary change even in a third situation, i.e. when the population is small. This condition was investigated by Sewall Wright (1921 and 1931) and led to the discovery of a second evolutionary mechanism.

In a large population, the sum of all random changes does not affect the mathematical centre of the genetic pool, since deviations occur in all directions with the same frequency and compensate each other; but in a small population this statistical levelling is not guaranteed, and the centre of the genetic pool makes an unpredictable shift at each generation. The trajectory described by that centre after many generations looks like the zigzag path of a drunk, or the erratic route of a raft that is going adrift, which explains why the process was given the name of *genetic drift*. The point is that a random walk is not likely to come back to the beginning, and the genetic pool undergoes therefore a permanent change, i.e. a true evolution is taking place, even if in a totally random way. Population genetics, in short, discovered that evolutionary change can be produced by three factors: (1) mutations, (2) natural selection and (3) genetic drift. Mutations are the only events that change hereditary characters, and must therefore be present in any evolutionary mechanism. This means that there are two distinct mechanisms of evolution: the mechanism of Darwin (mutation+natural selection) and the mechanism of Wright (mutation+genetic drift).

But does genetic drift exist in nature? Wright pointed out that an example had already been discovered by the reverend John Gulick in 1872. Gulick studied the *Achatinella* landsnails of the Hawaiian islands and found that the resident species of different valleys had developed very singular differences in their shells. Since the habitats were extremely similar, the changes could not be attributed to natural selection, and Gulick explained them as a result of variations that became fixed at random and were preserved by reproductive isolation. In 1888 Gulick published his results in a paper entitled "Divergent Evolution through Cumulative Segregation", and a year later Alfred Wallace himself recognized that Gulick's mechanism was indeed different from natural selection. Wright gave to genetic drift the name of *Gulick's effect*, but since it was he who proved its general nature and its mathematical basis, it has become known as *Sewall Wright's effect*. For our purposes, however, what matters is that genetic drift is not only predicted by theory but actually exists in life, and truly represents a second mechanism of evolution.

The Modern Synthesis

The rediscovery of Mendel's laws, the demonstration of crossing-over, the chromosome theory of heredity, the link between sex and XX or XY chromosomes, and the discovery of the first Mendelian disorders in man (alcaptonuria and brachydactyly) were all obtained in the first ten years of the twentieth century. In that brief period of time, light was thrown on the millennial mystery of heredity, and genetics became a science.

A change in a gene was called a *mutation*, a sober word for "sudden creation of hereditary novelty", and the first geneticists (Hugo de Vries, Carl Correns, William Bateson and Wilhelm Johannsen) regarded mutations as the only real moving power of evolution. They almost instinctively rejected natural selection, but this is understandable because they were studying the genetics of individual organisms, and in this field mutation is everything.

Later on, however, as attention shifted from individuals to populations and geneticists learned the lesson of the Hardy–Weinberg law, it became clear that mutations alone are not enough. People realised that an evolutionary mechanism must necessarily be a two-step process: a first stage where mutations occur, and a second one where mutations spread in a population and their destiny is actually decided (many are lost and only a few become "fixed"). The deciding force, as we have seen, can be either Darwin's natural selection or Wright's genetic drift, and this of course raises the problem of understanding which of the two is more important in nature.

Since adaptive characteristics can only be produced by natural selection and adaptation is by far the dominant aspect of life on Earth, there is no doubt that phenotypic evolution has largely been adaptive evolution, and has been shaped therefore by natural selection. Population genetics rediscovered in this way the key role of natural selection, and not only regarded it as vastly more important than genetic drift, but also gave it a formidable mathematical basis. The two biological fields were clearly converging and their unification became inevitable. The synthesis of genetics and natural selection – of Mendelism and Darwinism – was actually realised in the 1930s by Ronald Fisher, J.B.S. Haldane and Sewall Wright, and has become known as the *synthetic theory of evolution*, or the *Modern Synthesis*.

Later on, the key role of natural selection was also recognised in other fields, and the Modern Synthesis was enriched by a second confluence of disciplines. This extension was realised by various authors, in particular by Theodosius Dobzhansky (1937), Ernst Mayr (1942) and George Gaylord Simpson (1944). Dobzhansky oulined the importance of selection in experimental genetics, and Mayr in biogeography and systematics, while Simpson gave a particularly valuable contribution to the synthesis by arguing that natural selection is perfectly compatible with the fossil record, a conclusion that many paleontologists before him had repeatedly denied.

Simpson was aware that geological strata do not display that *"gradual succession of countless intermediate forms"* that Darwin expected, and knew only too well that the fossil record is full of discontinuities that suggest sudden changes, rather than continuous transformations. He pointed out, however, that geological time is quite different from biological time. A million years is only a geological instant, but corresponds to many thousands of generations in any species, and this is just the number that varieties need, on average, to accumulate the changes that transform them irreversibly into new species. Simpson concluded therefore that the gradual accumulation of small changes produced by natural selection is perfectly *compatible* with the discontinuities of paleontology.

He also pointed out that the gradual evolution of species (technically known as *phyletic gradualism*) can be achieved by two distinct processes (Figure 2.2): (1) a species can undergo a continuous transformation without increasing the number of species that live at any one time (*phyletic transformation*), and (2) an ancestral species can split into two or more groups of descendant species (*phyletic speciation*). This splitting is the only mechanism that can actually increase the diversity of life on Earth, and it is certain therefore that phyletic speciation has occurred during the history of life. Simpson argued, however, that phyletic transformation must also have taken place, because its continuous transformation of species can account, in principle, for the appearance of higher taxa. This allowed Simpson to conclude that *macroevolution* (the origin of taxa above the species level) can be explained by *microevolution* (the origin of species).

Phyletic gradualism (transformation and speciation) became in this way the one and only mechanism of evolution in the framework of the Modern Synthesis, and Simpson's contribution was welcome as the long-awaited reconciliation of natural selection with paleontology.

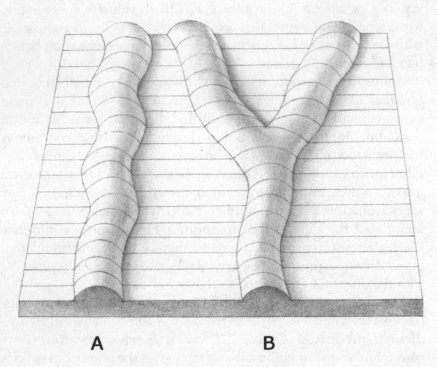

A   **B**

Figure 2.2   According to Darwin's theory of natural selection, species evolved by a mechanism of small continuous changes that today is known as *phyletic gradualism*. In this framework, evolutionary change can be realized by two distinct processes: (A) a gradual transformation of a species without any increase in the total number of species (*phyletic transformation*), and (B) a gradual separation of an ancestral species in two or more descendant species (*phyletic speciation*).

Molecular evolution

The laws of population genetics are usually expressed with technical terms such as *mutation frequency*, *nucleotide-substitution rate*, *amino-acid-turnover rate* and so on, but until the 1960s it was impossible to make direct measurements of these parameters. The only way to estimate them was by deducing their values from their visible effects on real organisms, i.e. by transferring to the molecular level what is observed in the phenotypic world.

A mutation, on the other hand, can not only be positive or negative for a given organism, but can also be *neutral*, in the sense that it could have no adaptive value. In this case natural selection would not work on it, and its destiny would be determined by the only other existing mechanism, i.e. by genetic drift. In order to understand evolution at the molecular level, therefore, it was necessary to estimate how many neutral mutations occur in nature, on average, compared to adaptive ones. Since direct measurements were impossible, it seemed logical to resort to indirect observations and to say that the ratio between neutral and adaptive mutations must be fairly close to the ratio between neutral and adaptive phenotypic characters. And since neutral characters are a tiny minority, it was concluded that neutral mutations must be a very small percentage of the total. This is why Ernst Mayr concluded, in 1963, that *"It is highly unlikely that really neutral genes do exist, or that a gene could remain neutral for a long time."*

Another way of obtaining indirect information on molecular evolution was offered by the study of phylogenetic trees. A typical example, in this field, is the comparison between amphibians and mammals. Both groups derived from a common aquatic ancestor, but amphibians evolved much more slowly. They share so many anatomical characters that a single order comprises most of them, while mammals differentiated into as many as sixteen distinct orders. Mammals clearly underwent a much faster phenotypic evolution than amphibians, and it seemed logical to conclude that, at the molecular level, *the mutation rate has been much faster in mammals than in amphibians.*

This and many other similar conclusions reinforced the idea that phenotypic evolution, molecular evolution and macroevolution all took place almost exclusively by natural selection and only minimally by genetic drift. This theoretical conclusion, known as *selectionism*, or more precisely *panselectionism*, became increasingly popular in the 1940s and 1950s, and up until the 1960s there was no serious alternative. In the 1960s, however, it became possible to make direct measurements of the molecular parameters, and the results turned out to be quite different from those predicted by selectionism.

It was found, for example, that mutations accumulate in the genomes of amphibians and mammals *at the same rate*. Phenotypic evolution occurred at very different rates in the two groups, but molecular evolution did not. It is not possible therefore to extrapolate from the phenotypic world to the molecular level, and the two levels must be kept apart. This creates of course an additional problem, because now we have to find a bridge between the two worlds, but it is nature itself that creates the problem and we can only do our best to solve it.

In addition to mutation rate, even the other molecular parameters turned out to be different from the expectations of selectionism. It was discovered, for example, that neutral mutations are not in the least a tiny minority with respect to adaptive mutations, and the actual ratio is probably the other way round. At the molecular level, in other words, the dominant mechanism of evolution is not natural selection but genetic drift, and this led Motoo Kimura to formulate *the neutral theory of molecular evolution* (1968, 1983).

This theory, in Kimura's own words, states that *"The great majority of evolutionary mutant substitutions are not caused by positive Darwinian selection but by random fixation of selectively neutral or nearly neutral mutants"* (it is important to underline that the adjective *neutral* does not mean *without function;* it only means that a mutation is adaptively indifferent, i.e. it is neither better nor worse than the previous one in respect to the organism's adaptation to the environment).

Kimura's theory has not been universally accepted, and the debate between selectionism and neutralism is still going on, but the experimental data have changed for good our view of molecular evolution. Today, biologists are aware that neutral mutations are a fact of life, and that genetic drift is, at the molecular level, at least as important, if not more important, than natural selection. It must also be noticed that this does not diminish in the least the key role of natural selection in *phenotypic* evolution, and Kimura himself explicitly acknowledged that *"The basic mechanism of adaptive evolution is without doubt natural selection."* He added however that *"Underneath the remarkable procession of life and indeed deep down*

at the level of the genetic material, an enormous amount of evolutionary
change has occurred, and is still occurring. What is remarkable, I think,
is that the overwhelming majority of such changes are not caused by
natural selection but by random fixation of selectively neutral or nearly
neutral mutants. Although such random processes are slow and
insignificant for our ephemeral existence, in the span of geological times
they become colossal. In this way, the footprints of time are evident in
all the genomes on the earth. This adds still more to the grandeur of our
view of biological evolution."

The third mechanism of evolution

The *chromosome theory of heredity* (proposed by Sutton and Boveri
in 1902) stated that genes are carried on chromosomes, but did not
say anything about their relative positions. It was Thomas Hunt
Morgan who picked up that problem. He could not build
chromosomal maps without an hypothesis on gene positions, and so
he chose the simplest: the idea that genes do not move around in
chromosomes but occupy *fixed* positions on them, like passengers
who are glued to their seats in train carriages. This is in fact what we
mean when we say that genes behave in a *Mendelian* way.

The very success of Morgan's maps gave an implicit endorsement
to the underlying hypothesis, and so the *fixity of gene position*, i.e.
the *Mendelian behaviour of genes*, became accepted as an article of
faith, so much so that the hypothesis was not even mentioned in the
Hardy–Weinberg famous theorem of population genetics. This
theorem, as we have seen, allows for only two exceptions to
evolutionary stasis, and therefore for only two mechanisms of
evolution (natural selection and neutral drift), but that was only
because the Mendelian behaviour of genes was taken for granted.
What if genes were to move around in chromosomes? Wouldn't
that mean that there is a third way out of the Hardy–Weinberg
prison, and therefore a third mechanism of evolution? Yes it would,
in principle, but in practice everybody's advice was to forget it. Again,
the very success of genetics was enough reason for not questioning

the foundational hypotheses of Mendelism.

Enough for *almost* everybody, that is. From the 1940s well into the 1960s, probably the only person in the world who had serious doubts was Barbara McClintock, and she wasn't exactly popular for that. In the 1940s, while studying the genes that affect the colour of kernels in maize, McClintock noticed that some white kernels were carrying scattered plum-coloured spots. In many groups of cells, in other words, one or more mutations had taken place which turned the colour from white to plum. The unexpected was not the mutations *per se*, but their astonishing frequency. The natural rate of mutation is about one in a million for any given gene, and yet McClintock was observing the same mutation taking place simultaneously in thousands of cells of the same plant in just one generation.

The only solution she could come up with (McClintock, 1951, 1956) was to assume that the mutations were propagated by genes jumping from one chromosome to another. There was enough for burning her at the stake for heresy, and in a figurative way that is what happened, as she herself recollected after receiving the Nobel Prize in 1983 (aged 81). The Nobel Prize came of course as a recognition that she had been right all along, a fact that became universally accepted only when mobile genes were discovered not only in plants but in all other forms of life, from microbes to animals.

And that was only the first of a series of discoveries which followed in quick succession and literally opened the gates of non-Mendelian heredity. Other phenomena – such as *unequal crossing-over*, *DNA slippage* and *gene conversion* – proved that the genome is actually a turbulent superstructure in which genes are in a continuous state of flux. The Mendelian behaviour of genes is only a crude approximation of the truth, good enough for many practical purposes but not for a real-life understanding of the *fluid genome*. This brings us back to the possibility of a third exception to the Hardy–Weinberg theorem, i.e. to the possible existence of a third mechanism of evolution based on non-Mendelian heredity. And since the new mechanism would be a direct result of gene turbulence, a good name for it could be *evolution by genomic flux*.

Since the early 1980s, Gabriel Dover has been one of the most outspoken supporters of the third mechanism, but has chosen a different name for it, and has spoken of *evolution by molecular drive* (Dover, 1982, 2000). In general, of course, it is not advisable to have two names for the same thing, but in this case the distinction does have a purpose, because biologists are divided on the issue. The majority view is that we do not yet know the actual evolutionary outcomes of non-Mendelian heredity, and so we must keep our options open. In this case we can indeed speak of *evolution by genomic flux* because the fluid genome is a reality, but we cannot be specific about its results. With the term *evolution by molecular drive*, instead, Dover refers to a mechanism which has a distinct outcome, and this is why it is proper to keep the two names separate.

Dover extended a proposal made by Stephen Jay Gould and Elizabeth Vrba (1982) on the need to use different words for different evolutionary processes. Ever since Darwin, biologists have only used the word *adaptation* for expressing what goes on between organisms and environment, but that does not cover the full range of possibilities. Neutral drift, for example, could produce features that organisms may "co-opt" for new purposes, and in that case the word adaptation would be misplaced. Neutral drift is non-adaptive by definition, and so Gould and Vrba proposed the word *exaptation* for that phenomenon. With a similar spirit, Dover pointed the finger to a different phenomenon that he called *adoptation*. He noticed that molecular drive could give some organisms the possibility of exploiting a new environment instead of adapting to the existing one. In this case, organisms would not "adapt" to an environment, but "adopt" one.

Dover underlined that there is no rigid one-to-one relationship between evolutionary mechanisms and environmental processes, but noticed nonetheless that a certain correspondence does exist. He concluded therefore that *adaptation* is the normal result of natural selection just as *exaptation* is characteristic of neutral drift, and *adoptation* is typical of molecular drive. To our purposes, what matters is that we can indeed speak of three mechanisms of evolution: the first two are *natural selection* and *neutral drift*, while the third mechanism can be called either *genomic flux* or *molecular drive*.

Macroevolution

In 1972, Niles Eldredge and Stephen Jay Gould published in *Models in Paleobiology* a paper whose title sounded like a war declaration: "Punctuated Equilibria: An Alternative to Phyletic Gradualism". As we have seen, *"phyletic gradualism"* is the name that was given to Darwin's classical gradualism by the proponents of the Modern Synthesis, and an *alternative* to this concept appeared to deliver a direct challenge to Darwinism, to natural selection and to the entire Synthesis. In reality, Eldredge and Gould had nothing of the kind in mind, and their paper was simply an attempt to show that macroevolution is a much more complex phenomenon than people had thought.

They started by pointing out that species appear *quickly* in the fossil record and then remain *unchanged* for long geological periods, that is for tens of millions of years (Figure 2.3). The fossil record is far from perfect, but this pattern is so frequent and regular that it cannot be dismissed as an exception. It is a veritable rule of the history of life, and must therefore be explained. Eldredge and Gould looked for a model that could account for that regularity, and found it in the idea of *allopatric speciation* proposed by Ernst Mayr (*allopatric* means *in another country* and allopatric speciation therefore is something that species do *somewhere else*).

"The central concept of allopatric speciation is that new species can arise only when a small local population becomes isolated at the margin of the geographic range of its parent species. Such local populations are termed 'peripheral isolates'. A peripheral isolate develops into a new species if 'isolating mechanisms' evolve that will prevent the re-initiation of gene flow if the new form re-encounters its ancestor at some future time. As a consequence of the allopatric theory, new fossil species do not originate in the place where their ancestors lived. It is extremely improbable that we shall be able to trace the gradual splitting of a lineage merely by following a certain species up through a local rock column".

Eldredge and Gould concluded that the history of species is not a uniform accumulation of small changes, but is made of long periods of stasis (the *equilibria*) which are occasionally interrupted by quick

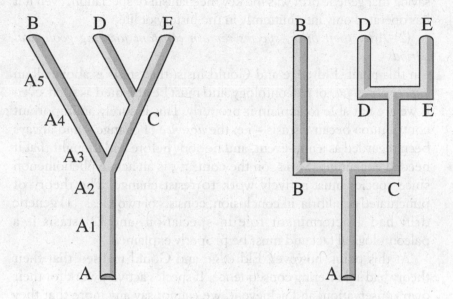

Figure 2.3   The basic patterns of species evolution according to phyletic gradualism (left) and punctuated equilibria (right). In the last model, horizontal lines indicate that speciations took place in geologically brief periods, and the fact that letters do not change along vertical lines signifies that species remained unchanged for long geological periods of time.

episodes of speciation (the *punctuations*). *Evolution is like the life of a soldier: long periods of boredom and short moments of terror.* But let us examine the central points of the theory.

(1) *Speciation occurs rapidly in peripheral isolates*
Eldredge and Gould declare that their model derives from Ernst Mayr's allopatric speciation, but this concept, in turn, was based on Sewall Wright's genetic drift. In small populations, as we have seen, genetic drift becomes the dominant mechanism of mutations' transport, and a few thousand generations are normally enough for the novelties to become fixed and for a population to undergo an irreversible change. The theory of punctuated equilibria amounts to

saying that genetic drift was the key mechanism of speciation, even if it has operated only intermittently in the history of life.

(2) *After their birth, species remain constant for long geological periods*

On this point, Eldredge and Gould insist that stasis is above all an *experimental fact* of paleontology, and must be accepted as such, even if we are not able to explain it properly. This is surely an important contribution because stasis – i.e. the *absence* of change – had always been regarded as a non-event, and nobody before had thought that it needed explaining. Stasis, on the contrary, is an active phenomenon since species must actively work to resist change. The theory of punctuated equilibria, in conclusion, consists of two ideas: (1) genetic drift had a determinant role in speciation, and (2) stasis is a paleontological fact and must be properly explained.

At this point, however, Eldredge and Gould realised that their theory had a shattering consequence. If species actively work for their own conservation, and achieve it, we cannot say any more that they are continuously transforming themselves in order to express the features of higher taxa. This is precisely the mechanism invoked by phyletic gradualism in order to explain macroevolution, but we cannot have it both ways: stasis and phyletic gradualism are not compatible, and cannot both be true. And since it is stasis that is documented by the fossil record, Eldredge and Gould concluded that phyletic gradualism must be abandoned, even if this means that we no longer have a model for macroevolution.

The question, at this point, was *que faire?* When a widely accepted solution is found wanting, the need to replace it with an alternative model is strong, and Eldredge and Gould discovered that a way out is offered by the idea of *species selection.* If species are regarded as organic systems that are born, live and die as individuals do, one can conclude that a selection mechanism operates at the species level as it does at the individual level.

This solution kills the proverbial two birds with one stone. On the one hand, it is acknowledged that microevolution and macroevolution are two separate hierarchical levels, and therefore that each of them has its own independent laws. On the other hand, it is said that one level is ruled by the natural selection of organisms and the other by the natural selection of species, which means that all problems are solved by natural selection. The explanation is ingenious, no doubt about that. But is it true? Unfortunately we have no proof of this, and the mechanism of macroevolution remains therefore a mystery.

Where is biology going to?

The various fields that make up a science tend naturally to integrate, and with time such a process can lead to a true synthesis. Physics was the first science to achieve a synthesis of its disciplines, and it may be useful to compare that experience with its biological counterpart. The first unification occurred between mechanics and thermodynamics, in the first half of the nineteenth century, and the second came shortly afterwards, with the integration of electromagnetism. The result was the imposing edifice of classical physics, a conceptual system that described all reality in terms of particles and waves, with equations that seemed perfect because they were perfectly deterministic. The common denominator of all branches of classical physics was in fact the concept of *determinism*, and nodoby doubted, in the nineteenth century, that that was the true logic of the universe.

In the first years of 1900, however, physicists started studying the structure of atoms and soon realised that it was impossible to describe them with models of either particles or waves. Those models were inevitable in classical physics, and their failure could only mean that that physics is not valid at the atomic scale. The laws which apply to one level of reality are not necessarily valid at other levels, but this did not stop the unification process. In the end, quantum mechanics did manage to account for the atomic world, and it turned out that it could also explain the results of classical physics, which means that a bridge can actually be built between different levels of reality. A new synthesis, in other words, became possible because the quantum description of nature was able to contain, as a particular case, the description of classical physics.

In biology, as we have seen, the Modern Synthesis started with the unification of natural selection with genetics, and a second step came with the addition of biogeography, systematics and paleontology. There is a clear parallel between this unification and that of classical physics, because both syntheses have in common the idea that only one logic applies to all levels. The authors of the Modern Synthesis, in fact, were also the fathers of panselectionism, and repeatedly stated that natural selection shapes all levels of life.

At this point, however, it must be underlined that physics and biology went through two very different historical debates. While the crisis of classical physics was followed by a collective attempt to build a *more advanced* physics, in biology any imperfection of the Modern Synthesis was used by many scholars as a pretext to return to *more backward* positions, i.e. to vitalistic or finalistic principles. Various attempts had been made, before Darwin, at explaining evolution with vital forces, with oriented tendencies, or with transcendental laws of form, and these ideas have been repeatedly reproposed, after Darwin, with other names (orthogenesis, hologenesis, noogenesis and so on). In that context, the defence of selectionism appeared to many as the defence of rationality itself in biology, and this does justify it.

Today we are, luckily, in a different position. Thanks to Kimura's work it is clear that molecular evolution and phenotypic evolution are not driven by the same mechanism. And thanks to Eldredge and Gould we have become aware that macroevolution is not easily reducible to microevolution. In biology, therefore, we are where physics was after the discovery that the laws of our middle-dimensional world (the *mesocosm*) are not necessarily valid at very small dimensions (the *microcosm*), or at very large ones (the *macrocosm*). We have not yet reached, however, the stage of theoretical synthesis that physics obtained with quantum mechanics, because we do not have a link between the various levels of life, and in particular because we do not have a real unification of evolution and embryology. This tells us that the goal of a true biological synthesis is *discovering the bridge that exists between genes and organism in order to understand how molecular evolution, phenotypic evolution and macroevolution are linked together.*

At this point, however, the history of physics reminds us that quantum mechanics required, as a *conditio sine qua non*, the rejection of the "logic" of classical physics. This may well be a general prerequisite, and biologists should be prepared to do with *selectionism* what physicists did with *determinism*. Selectionism of course would still be used in practice, exactly as Newton's laws are still used in middle-dimensional mechanics, but it would no longer be the deep "logic" of life. While firmly rejecting any return to backward positions, in other words, we should be prepared to accept the idea that *no real unification in biology is possible without a new logic of life*. The search for that logic becomes therefore our goal, and the rest of this book is devoted to show that today we can make at least a few steps into that new unexplored world.

3

A NEW MODEL FOR BIOLOGY

This chapter describes a mathematical model of epigenesis, and starts by translating the traditional but vague definitions of that concept into expressions that are increasingly more precise. The first step consists in defining epigenesis as *a convergent increase of complexity*. The second states that this process is equivalent to *a reconstruction from incomplete information*, and in the third step this becomes *a reconstruction from incomplete projections*. In this way, we can model epigenesis as a special case of the problem of reconstructing structures from projections, a problem that arises in many fields (for example in computerised tomography) and whose mathematics is well known. What is less well known is that a reconstruction can be achieved even when the starting information is incomplete, provided that appropriate memories and codes are employed. This is illustrated with a few practical examples, and the logic of that unusual kind of reconstruction is described first in words and then in formulae. At that stage we can go back to biological epigenesis and conclude that a convergent increase of complexity in organic life necessarily requires organic codes and organic memories. And this gives us the two critical concepts that will be used in the rest of the book for an entirely new approach to the problem of biological complexity.

The logic of embryonic development

The discovery of genes that control embryonic development has started a true revolution in biology, both from an experimental and from a theoretical point of view. On the experimental side, it has opened fields of research that previously seemed unapproachable.

From a theoretical point of view, it has inspired the conclusion that embryonic development is the execution of a genetic program, in the sense that all processes of ontogenesis depend, more or less indirectly, on the transcription of genes. Unfortunately, many have also concluded that the central problem of development – the problem of form – has been, *in principle*, resolved. Many details are still to be worked out, it is said, but the "logic" is now clear because the form of an organism depends on its genes.

In his book *The Problems of Biology* (1986), John Maynard Smith has lucidly sounded a note of caution against this attitude:

"It is popular nowadays to say that morphogenesis (that is the development of form) is programmed by the genes. I think that this statement, although in a sense true, is unhelpful. Unless we understand how the program works, the statement gives us a false impression that we understand something when we do not ... One reason why we find it so hard to understand the development of form may be that we do not make machines that develop: often we understand biological phenomena only when we have invented machines with similar properties ... and we do not make 'embryo' machines."

Maynard Smith's point can also be expressed in another way: embryonic development is a process that increases the complexity of a living system, but we do not know how to build machines that increase their own complexity, and we cannot therefore understand the logic of development. We can also leave aside the physical construction of machines and concern ourselves only with their planning. If we could prove, with a mathematical model, that it is possible to increase the complexity of a system, we already would have taken a major step forward. The search for the logic of development begins therefore with the search of a mathematical model for systems which are capable of increasing their own complexity.

At this point, however, a formal distinction between two very different cases is called for. An increase in complexity took place even during the history of life, but in this case new structures arose by chance mutations, and the increase was a *divergent* process. In embryonic development, on the contrary, new structures are never formed by chance, and we are dealing with a *convergent* increase of complexity. This is the great difference between evolution and ontogenesis, and such a dichotomy does require two very different types of mathematical models.

In the case of evolution, we already have algorithms that simulate the effects of natural selection, and we do therefore understand how a divergent increase of complexity can take place. But we do not have algorithms that describe a convergent increase, and it is for this reason that the logic of the embryos still eludes us. The real key to embryonic development is the logic of systems which are capable of increasing their complexity *in a convergent way*, and in order to understand this we need, if not a machine, at least a model that is functioning according to that logic.

Reconstruction from incomplete projections

The starting-point for a new model of embryonic development is the reconstruction of structures from their projections, a problem which arises in many fields such as computerised tomography and electron microscopy. The image produced by X-rays on a radiographic plate, for example, is a projection of a three-dimensional body on a two-dimensional surface, and this process is inevitably accompanied by a loss of information. The result, to quote Hounsfield (1972), is *"like having a whole book projected on a single sheet of paper, so that the information of any one page cannot be extracted from the superimposed information of all the other pages"*. In order to reconstruct the original structure, therefore, it is necessary to collect a plurality of projections at different angles, as shown in Figure 3.1. The minimum number of projections that must be collected is known from basic theorems, and has an clear intuitive meaning. The projections taken at different angles carry different information, and their totality must contain (in a compressed form) all the information that was present in the original structure.

As for the reconstruction algorithms, we can divide them in two major groups: iterative and non-iterative techniques. A non-iterative method produces the final result with a formula which is applied only

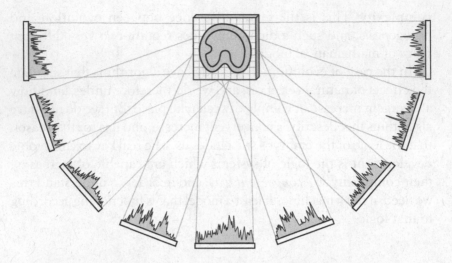

Figure 3.1   During the projection of a three-dimensional structure onto a two-dimensional plane, information is lost, and it is therefore necessary to collect a plurality of projections at different angles in order to reconstruct the original structure.

once to the experimental data. In this case the reconstruction is precise, because the formula provides a rigorous solution, but the procedure is cumbersome because all data are processed together. The iterative algorithms have been introduced precisely in order to simplify the reconstruction procedure and still obtain satisfactory results. In these cases, a reconstruction produces only an approximation of the original structure, and it is therefore necessary to repeat the operations many times in order to get progressively closer to the original structure.

Iterative algorithms are clearly less precise than single-application techniques, but their great advantage is that they introduce the *time dimension* in the computation, and this makes them particularly suitable to simulate biological processes. Even more important is the fact that the temporal dimension allows us to reconsider the problem of the minimum number of projections that are required for a complete reconstruction. During an iterative procedure, we could discover properties of the original structures that were not recorded in the projections, and in this case we could obtain a reconstruction even if the number of projections is appreciably lower than the theoretical minimum. This brings us face to face with an entirely new problem: *the problem of reconstructing structures from incomplete projections*, where projections are said to be *incomplete* when their number is at least one order of magnitude less than the theoretical minimum. The problem, in other words, is to make a complete reconstruction with an amount of information which is much lower than that of the original structure.

The interesting point is that this is a mathematical version of the problem that we face in embryonic development. The fertilised egg contains far less information than the adult organism (whatever criterion is used to measure information in biological systems), and embryonic development can be described therefore as a process that is reconstructing a structure from *incomplete information*. This is another way of saying that embryonic development is a process that increases the complexity of a living system. The reconstruction of structures from incomplete information, in short, is a model that could help us understand how it is possible for a system to obtain a convergent increase of complexity.

A memory-building approach

An iterative reconstruction algorithm produces a series of pictures which are increasingly more accurate approximations of the original structure. Any reconstructed picture is affected by errors, and in general there is no way of knowing where the errors are falling, but there are two outstanding exceptions to this rule. The values which are below the minimum or above the maximum are clearly "illegal", and the algorithm gives us the precise coordinates of the points where they appear. This makes it possible to correct those errors by setting to the minimum or to the maximum all values which are respectively below or above the legal limits. This operation (which is called a *reconstruction constraint*) does improve the results and so it is normally applied at regular intervals. Every time that we apply this constraint, however, we lose some information, because the coordinates of the illegal values are lost. Keeping information about errors may not seem important, but let us assume that it could be, and let us see if we can save it. This can be done by using a "memory" picture where we store the coordinates and the values of the illegal points before applying the constraint. In this case we need a more complex algorithm, because we have to perform in parallel two different reconstruction: one for the structure and one for the memory, as illustrated in Figure 3.2. But what is the point of keeping a memory of the reconstruction errors?

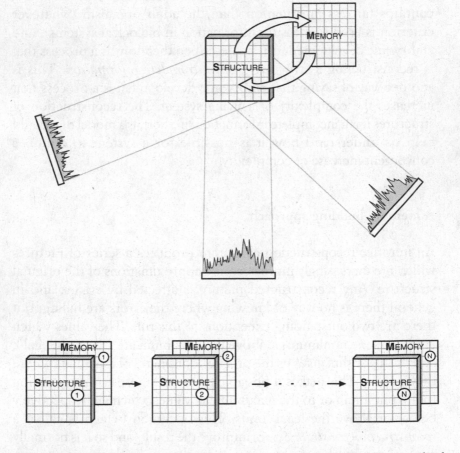

Figure 3.2   A reconstruction from incomplete projections is a method where structure matrices and memory matrices are reconstructed in parallel.

The point is that we can study their pattern, and that turns out to have unexpected features. Since the errors are random events, we would expect a statistical distribution, but this is not what happens in all cases. It is true that in many points the errors are totally random, but there invariably are other points where this does not happen. In those points the illegal values keep reappearing each time, and always with the same sign, which explains why such points have been called *vortices*. Figure 3.3 is a schematic illustration of what happens. The patterns of the illegal values look totally random when they are examined one by one (Figure 3.3A), but when they are memorised together (Figure 3.3B) the statistical fluctuations disappear and only the vortices stand up. Now we have a new type of information before us. When an illegal value has consistently reappeared in the same point for a number of times (we can choose 5, 10 or any other convenient number), we can reasonably conclude that the value of that point is either a minimum or a maximum. We can therefore "fix" the value of that point, and this means that the total number of the unknowns is reduced by one. By repeating the operation, the number of the unknowns becomes progressively smaller, and when it reaches the number of the equations a complete reconstruction is possible. That is the result we were looking for. With appropriate "tools" we can indeed obtain a complete reconstruction of the original structure from incomplete information.

Let us now take a closer look at those "tools". One is the memory picture that must be reconstructed in parallel with the structure, but that is not all. The new information of the vortices appear in the memory space, but we use that information in the structure space, because it is here that we reduce the number of the unknowns. We are in fact transferring information from the memory space to the structure space with a *conventional rule* of the type *"If a vortex appears in the memory space, fix the corresponding point in the structure space to a minimum or a maximum."* A reconstruction from incomplete information, in short, does not require only a memory. It requires memory and codes. The reconstruction memory is where new information appears. The reconstruction codes are the tools that transfer information from the memory to the structure.

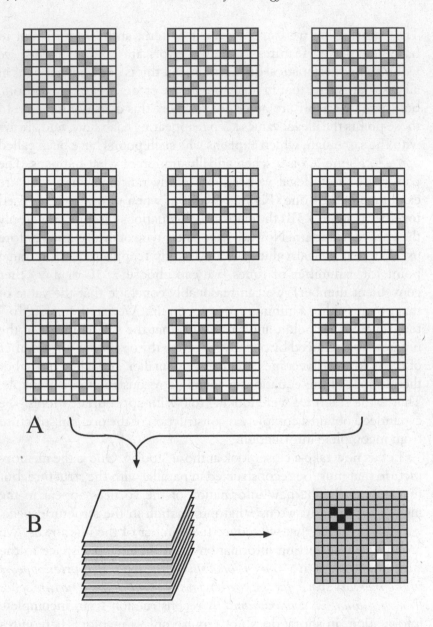

Figure 3.3   The errors produced by an iterative reconstruction algorithm have patterns which appear, at each iteration, completely random (A), but if successive patterns are memorised together, it is possible to observe regular structures appearing in the memory matrix (B).

The biological implications of the above model are straightforward. Embryonic development is also a reconstruction of structures from incomplete information, and so it must employ organic memories and organic codes. Before looking for the presence of these biological tools in nature, however, we must examine the mathematics of the new reconstruction method. Today there is a widespread belief that a convergent increase of information is impossible (despite the evidence from the embryos) and only mathematics can give us the proof of the contrary.

The algebraic method

The simplest case is the reconstruction of two-dimensional structures from one-dimensional projections. A digitised two-dimensional structure, for example a television picture, can be described as an $n \cdot n$ matrix $[f_{ij}]$ of side D and cells (i,j) of side $d = D/n$ (Figure 3.4). A projection of the picture at an angle ϑ is a set of parallel rays (ϑ,k) which totally cover the picture at the angle ϑ, and any projection ray can be represented by an $n \cdot n$ matrix (Figure 3.5) where each element $a_{ij}^{\vartheta k}$ is the fraction of the cell (i,j) which is contained within the ray (ϑ,k). The picture matrix and the ray-matrices are easily transformed into vectors (Figure 3.6). More precisely, the picture matrix $[f_{ij}]$ is replaced by a column-vector $[f_z]$, and the ray matrices $[a_{ij}^{\vartheta k}]$ are described by row-vectors $[a_z^{\vartheta k}]$ with the transformations:

$$f_{ij} \longrightarrow f_z \quad \text{and} \quad a_{ij}^{\vartheta k} \longrightarrow a_z^{\vartheta k} \quad \text{with} \quad z = 1, \ldots \ldots, n^2 = t$$

In this way, the projection values $g_{\vartheta k}$ of any ray (ϑ,k) are described by the scalar product of the vectors $[a_z^{\vartheta k}]$ and $[f_z]$:

$$g_{\vartheta k} = a_1^{\vartheta k} f_1 + a_2^{\vartheta k} f_2 + \ldots \ldots + a_t^{\vartheta k} f_t$$

which is a linear equation with $t = n^2$ unknowns.

If we have p projections of a picture and each projection contains r rays, we have a system of $p \cdot r$ equations in n^2 unknowns, and a solution

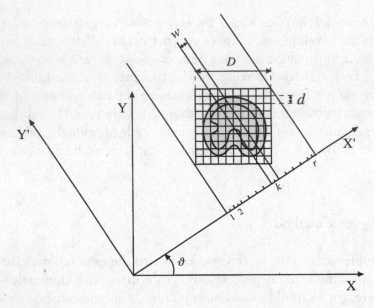

Figure 3.4   A digitized two-dimensional structure can be represented by a matrix of side *D* which is made of $n \cdot n$ cells, or pixels, of side *d*. The projection of a picture can be represented by a set of adjacent parallel rays, of equal width *w*, which totally covers the matrix of the picture.

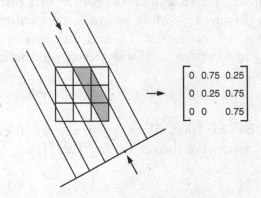

Figure 3.5   A projection ray that crosses an $n \cdot n$ picture matrix can also be represented by an $n \cdot n$ matrix, where each cell (i, j) of the ray matrix contains a number that represents the fraction of the cell (i, j) of the picture matrix which is contained within the projection ray.

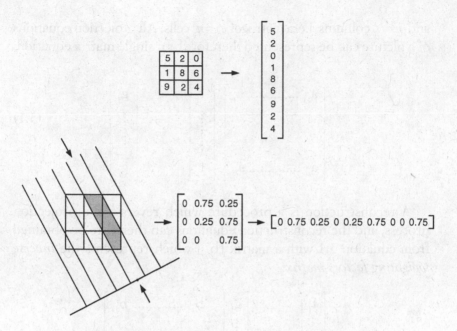

Figure 3.6   A picture matrix and a ray matrix can both be represented by vectors. More precisely, by a column vector for the picture matrix, and by a row vector for the ray matrix.

exists if the number of linearly independent equations is equal to the number of unknowns, i.e. if

$$p \cdot r = n^2$$

In order to have all equations in a compact form, the double index (ϑ, k) of each ray is replaced by the single index (h) with the transformations

$$g_{\vartheta k} \longrightarrow g_h \quad \text{and} \quad a_z^{\vartheta k} \longrightarrow a_{hz} \quad \text{with } h = 1, \ldots \ldots , p \cdot r = t$$

In this way, all the projections of a picture are represented by a single column vector $[g_h]$, and the geometrical parameters form a matrix $[a_{hz}]$, known as the *weighting factors matrix*, which has $p \cdot r = t$ rows and $n^2 = t$ columns, i.e. a matrix of $t^2 = n^4$ cells. All projection equations of a picture can be represented therefore by a single matrix equation:

$$
\begin{bmatrix} a_{11} \dots\dots\dots a_{1t} \\ \\ \\ \\ a_{t1} \dots\dots\dots a_{tt} \end{bmatrix}
\begin{bmatrix} f_1 \\ \vdots \\ \vdots \\ f_t \end{bmatrix}
=
\begin{bmatrix} g_1 \\ \vdots \\ \vdots \\ g_t \end{bmatrix}
\tag{3.1}
$$

A reconstruction is a procedure which reverses the projection process, and the reconstruction equations can therefore be obtained from equation 3.1 with a matrix $[b_{hz}]$ which represents the *inverse weighting factors matrix*:

$$
\begin{bmatrix} f_1 \\ \vdots \\ \vdots \\ f_t \end{bmatrix}
=
\begin{bmatrix} g_1 \\ \vdots \\ \vdots \\ g_t \end{bmatrix}
\begin{bmatrix} b_{11} \dots\dots\dots b_{1t} \\ \\ \\ b_{t1} \dots\dots\dots b_{tt} \end{bmatrix}
\tag{3.2}
$$

The values f_z of the reconstructed picture are obtained therefore by the following equations:

$$
f_1 = g_1 b_{11} + g_2 b_{12} + \dots\dots + g_t b_{1t}
$$
$$
\vdots
$$
$$
f_t = g_1 b_{t1} + g_2 b_{t2} + \dots\dots + g_t b_{tt}
\tag{3.3}
$$

Once the weighting factors are calculated, the reconstruction values are obtained by equations 3.3 with simple additions and multiplications. This classic algebraic method, known as *matrix inversion*, is rigorous and straightforward, but in practice it is employed only with small pictures because the weighting factors matrix contains n^4 cells, and its dimensions become quickly prohibitive with increasing values of n (for a picture with $100 \cdot 100$ cells we would need a weighting factors matrix with $100^4 = 10^8$ cells).

The theoretical limit

Matrix inversion is not widely used in practice, but from a theoretical point of view is extremely useful, because it allows us to calculate the minimum number of projections that are required for a complete reconstruction. If we have p projections of a structure, and each projection contains r rays, a reconstruction procedure amounts to solving a system of $p \cdot r$ equations in n^2 unknowns, and algebra tells us that a solution exists only if the number of *linearly independent* equations is equal to the number of the unknowns.

The condition that equations are linearly independent is easily understandable, because it amounts to saying that projections obtained at different angles must transport different information (if they didn't, the total information of the projections would be inferior to that of the original picture and the reconstruction would be impossible). In practice, the linear independence condition implies that (1) the angle between any two projections must be greater than a critical minimum (which means that the projections must be equally distributed in the 180° angular range), and (2) the ray width (w) and the cell width (d) must have the same order of magnitude, i.e.

$$w \approx d \tag{3.4}$$

Since $d = \dfrac{D}{n}$ and $w = \dfrac{D(\vartheta)}{r} \approx \dfrac{D}{r}$

equation 3.4 is equivalent to $\dfrac{D}{n} \approx \dfrac{D}{r}$ and therefore $n \approx r$.

This means that the requirement $p \cdot r = n^2$ becomes $p \cdot n \approx n^2$, which amounts to

$$p \approx n \qquad (3.5)$$

The result is that *the minimum number of projections that are required for reconstructing a structure of* n^2 *unknowns is comparable to the square root of the number of the unknowns.*

It is important to notice that, in real-life applications, the actual number of projections must always be *greater* (often much greater) than the theoretical minimum, because of the need to compensate the inevitable loss of information which is produced by various types of noise. It is also important to notice that the theoretical minimum obtained with *non-algebraic* methods (Crowther *et al.*, 1970) is never inferior to the algebraic minimum. Equation 3.5, in other words, is the *lowest possible* estimate of the minimum number of projections that are required for a complete reconstruction of any given structure.

ART: an iterative algebraic method

The first algebraic reconstruction method was described by Hounsfield in 1969 in a patent application for computerised tomography, and an equivalent version was published independently by Gordon, Bender and Herman in 1970 with the name of *ART (Algebraic Reconstruction Technique)*. Instead of resorting to the matrix inversion approach (which requires matrices of n^4 cells), the reconstructions of this iterative method are performed with matrices of n^2 cells, and are therefore much simpler to handle. The algorithm starts with a uniform matrix [f_{ij}^0 = constant], and performs an iterative sequence of corrections which tend to bring the reconstructed matrix increasingly closer to the original structure. The corrections consist in calculating the differences between the projection values of the original structure ($g_{\vartheta k}$) and those of the matrix reconstructed at iteration q, ($g_{\vartheta k}^q$), and then in redistributing these differences among the cells of the reconstruction matrix. The reconstruction values at iteration $q+1$ are obtained therefore from those of iteration q with the algorithm

$$f_{ij}^{q+1} = f_{ij}^q + \frac{g_{\vartheta k} - g_{\vartheta k}^q}{N_{\vartheta k}} \qquad (3.6)$$

where $N_{\vartheta k}$ is the number of cells whose central points (i,j) are inside the ray (ϑ,k).

Gordon and Herman (1974) gave to equation 3.6 the name of *"unconstrained ART"*, and called *"partially constrained ART"* the same algorithm subjected to the constraint that negative values are set to zero at the end of each iteration. In addition to this, they called *"totally constrained ART"* the version where negative values are set to zero and values which exceed the maximum M are set to M. At first, it may appear that the totally constrained algorithm requires an a priori knowledge of the maximum M, but in practice it is always possible to obtain a satisfactory estimate of M even without that information. This can be achieved with some preliminary runs of *unconstrained ART* and *partially constrained ART*, because it can be shown that the maxima M^u and M^p obtained with these algorithms satisfy the relationship

$$M^u \leq M \leq M^p$$

and an average of M^u and M^p gives an estimate of M which becomes increasingly accurate as the number of iterations increases.

Gordon and Herman have also proposed a variety of formulae which allow one to compute the distance between the original picture and the reconstructed matrix, and therefore to evaluate the efficiency of a reconstruction algorithm. The ART method, in conclusion, is simple, fast and versatile, which explains why it has become an ideal starting-point for research on a new class of reconstruction algorithms.

The memory matrix

In reconstructions performed with iterative algorithms we usually find, at each iteration, values that are below the minimum and above the maximum, but we have already seen that it is always possible to bring these "illegal" values within the legitimate range. Let us assume, however, that we want to discover something else about those irregular values, apart from the fact that they do exist. It could be interesting, for example, to find out whether their distribution in space is totally random or is following some kind of regularity.

In order to answer this kind of questions, we can perform reconstructions by using not only the structure matrix $[f_{ij}]$ but also an additional matrix $[m_{ij}]$, of the same size, where we "memorise" the illegal values which appear at each iteration. This allows us to conserve a "memory" of them even when they have been erased from the structure matrix, and for this reason their matrix has been called the *memory matrix.*

The construction of the memory matrix is performed by taking as a starting point a totally "blank" matrix $[m_{ij}^0 = 0]$, and by applying the following operations:

$$
\begin{aligned}
&\text{If } f_{ij} \leq 0 &&\longrightarrow && m_{ij} = m_{ij} - \gamma \\
&\text{If } f_{ij} \geq M &&\longrightarrow && m_{ij} = m_{ij} + \gamma \\
&\text{otherwise} &&\longrightarrow && m_{ij} = m_{ij}
\end{aligned}
\qquad (3.7)
$$

where γ is a parameter which is chosen to represent the presence of an "illegality" in any convenient way.

The combination of a totally constrained algorithm with equations 3.7 of the memory matrix allows us to build, at each iteration, two very different matrices: the structure matrix where the reconstruction appears, and the memory matrix where the parameters of the illegal values are gradually accumulated.

If the distribution of these values were totally random, the memory matrix would tend to remain uniform, but in reality its behaviour is much more complex than that. At many points the illegal values do have a random behaviour, in the sense that they appear and disappear in a statistical way, but at other points the illegalities keep reappearing with absolute regularity at each iteration, and always with the same sign. These points clearly behave as "attractors" of density, and for this reason have been called *vortices*. More precisely, the names *negative vortices* and *positive vortices* have been given to the points (or cells) where values appear which are respectively smaller than the minimum and greater than the maximum for T consecutive iterations (where T is a parameter which is chosen by the operator).

By indicating with V_0 the negative vortices and with V_M the positive ones, the recognition of the vortices is performed, every T iterations, with the following criteria:

$$
\begin{aligned}
\text{If } m_{ij} = -T\gamma & \longrightarrow & m_{ij} = V_0 \\
\text{If } m_{ij} = +T\gamma & \longrightarrow & m_{ij} = V_M \\
\text{otherwise} & \longrightarrow & m_{ij} = m_{ij}
\end{aligned}
\tag{3.8}
$$

Another important result is obtained by applying this method to pictures of many different kinds, because it has been noticed that the space distribution of the vortices is *picture-dependent*. The vortices' pattern does not depend therefore on general characteristics of the algorithm, but on specific properties of the examined picture. It is as if a picture had a specific image in the memory space exactly as it has one in the real space. This brings us immediately to the following question: *Is it possible to use the information that appears in the memory matrix to improve the reconstruction in the structure matrix?*

The question is absolutely natural because the vortices appear to have a precise, and often even obvious, meaning. If a negative (or a positive) density value keeps reappearing in the same point for T consecutive times, it is clear that in the original structure that point must be a minimum (or a maximum). But if this is true, it is clearly useless to keep treating that point as an *unknown*, and we can therefore erase it from the list of the unknowns. The advantage of this operation is obvious: *while the number of equations (p·r) remains constant, the number of the unknowns (n²) is decreasing.*

If this is confirmed, the problem of reconstructing structures from *incomplete* projections could be solved. The key obstacle, in this problem, is precisely the fact that the number of equations is smaller than the number of unknowns, but if the unknowns are continuously reduced, eventually they would reach the same number as the equations, and at that point an exact reconstruction would be guaranteed. As we can see, the production of "illegal" density values – which was looking like a structural defect of the algorithm – opens the way to unexpected developments.

Density modulation

The first algorithm to use memory matrices was presented at Brookhaven's first international workshop on reconstruction techniques with the name of *"density modulation"* (Barbieri, 1974a). This method recognizes the vortices with equations 3.7 and 3.8, and then subtracts them from the list of the unknowns. By indicating with $N^0_{\vartheta k}$ and $N^M_{\vartheta k}$ the number of negative and positive vortices that fall in the ray (ϑ, k), the values of the reconstructed matrix at iteration $q+1$ are obtained with the following instructions:

$$\text{If} \quad m_{ij} = V_0 \text{ or } V_M \quad \longrightarrow \quad f_{ij}^{\,q+1} = f_{ij}^{\,q}$$

$$\tag{3.9}$$

$$\text{otherwise} \quad f_{ij}^{\,q+1} = f_{ij}^{\,q} + \frac{g_{\vartheta k} - g_{\vartheta k}^{\,q}}{N_{\vartheta k} - N^0_{\vartheta k} - N^M_{\vartheta k}}$$

The results obtained with density modulation depend, as we have seen, upon the choice of a parameter T that represents how many times an illegal value must appear in a cell in order to be considered a vortex. If $T = 10$, for example, it is reasonable to conclude that the point in question is a true vortex, but in this case the procedure is lengthy and the number of unknowns decreases very slowly. The choice of $T = 5$, on the other hand, increases the speed of the algorithm but also increases the probability of making mistakes in vortex recognition.

The first reconstructions performed with density modulation were made with the choice $T = 5$, and the results (Figure 3.7D) clearly showed that some points had been erroneously classified as vortices. Despite these mistakes, however, the reconstructions obtained with density modulation were greatly superior to those of the other algorithms (Figure 3.7B and 3.7C), and the memory method therefore is effective even when the choice of its parameters is not ideal. The most important result, however, is another one. The original pictures (Figure 3.7A)

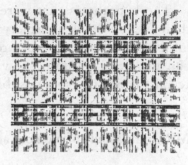

Figure 3.7 A black-and-white picture (A) reconstructed from 12 projections with Convolution (B), ART (C) and density modulation (D). The original picture was a 120 · 120 matrix, and in order to perform a complete reconstruction it would have been necessary to work with 120 projections in a full 180° angular range.

were matrices with 120·120 cells, and we know, from equation 3.5, that a complete reconstruction requires a minimum of 120 projections. The reconstructions of Figure 3.7 were made instead with only 12 projections, i.e. only 10% of the minimum information was actually used. This is clearly an example of *reconstruction from incomplete projections*. The results obtained with $T = 5$ would surely have been better with $T = 10$, but this is not the point. What really matters is that the main goal has been achieved even with $T = 5$. That goal was the proof that *the memory matrix does allow us to decrease the number of unknowns*, and the results tell us that this is precisely what happens.

The hypotheses that were made about density modulation, therefore, are valid: a memory matrix does allow us to obtain new information about the structure that we are reconstructing, and we can progressively move towards the point where a complete reconstruction becomes possible.

MRM: the family of memory algorithms

One of the interesting features of density modulation is that the reconstructions of black-and-white pictures (Figure 3.7) contain fewer errors than those obtained with grey (or *chiaroscuro*) pictures, i.e. with pictures which have intermediate degrees of density (Figure 3.8). This is understandable, because in black-and-white images all points are either minima or maxima, and the number of vortices is potentially very high. In grey pictures, instead, minima and maxima are far less numerous, and therefore the number of points that can be taken away from the list of the unknowns is much smaller.

This result is interesting because it focuses our attention on the *individual* features of the memory matrix. If only vortices are memorized, it is obvious that the algorithm performs better with pictures that have a high potential number of vortices, but if *other* features could be memorized, it would become possible to reduce substantially the unknowns even with grey pictures. We have therefore the problem of discovering if other features exist which allow us to

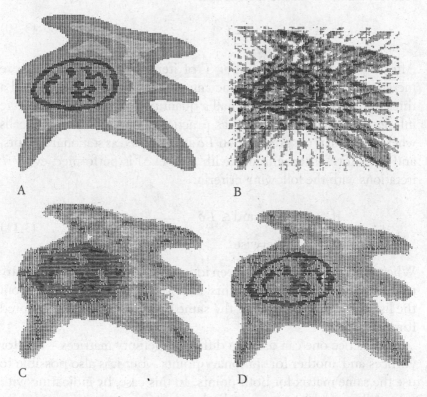

Figure 3.8   A grey, or *chiaroscuro*, picture (A) reconstructed from 12 projections with Convolution (B), ART (C) and density modulation (D). As in the previous case, a complete reconstruction would have required 120 projections equally spaced in the 180° range.

reduce the number of the unknowns, i.e. if there are other types of memory matrices.

A first hint came from the discovery that, in some cells, the reconstructed values can remain virtually unchanged for many consecutive iterations. In order to find these cells – which are called *stationary points* – it is necessary to keep a record of the values obtained in any two consecutive iterations, and to store their differences in a memory matrix with the instruction

$$m_{ij} = m_{ij} + |f_{ij}^{q+1} - f_{ij}^{q}| \tag{3.10}$$

After a predetermined number T of iterations, the sum of these increments is evaluated, and one can see whether it has exceeded a threshold $T\delta$, where δ is so small a quantity that any density change inferior to it can be regarded as practically insignificant. The cells where that sum is not greater than $T\delta$ are regarded as stationary points, and their formal recognition (with a label S) is performed every T iterations with the following criteria:

$$\text{If} \quad m_{ij} > 0 \text{ and} \leq T\delta \quad \longrightarrow \quad m_{ij} = S$$
$$\text{otherwise} \quad \longrightarrow \quad m_{ij} = m_{ij} \tag{3.11}$$

When stationary points have been identified, it is no longer necessary to treat them like the other points, and we can take them away from the list of the unknowns with the same procedure that was adopted for the vortices.

In practice one can use two different memory matrices – one for vortices and another for stationary points – but it is also possible to use the same matrix for both points. In this case, by indicating with $N^{0}_{\vartheta k}$, $N^{M}_{\vartheta k}$ and $N^{S}_{\vartheta k}$ respectively the negative vortices, the positive vortices and the stationary points that fall within the ray (ϑ, k), the values of the reconstruction matrix at iteration $q + 1$ are calculated with the following algorithm:

$$\text{If} \quad m_{ij} = V_{0}, V_{M} \text{ or } S \quad \longrightarrow \quad f_{ij}^{q+1} = f_{ij}^{q}$$

$$\tag{3.12}$$

$$\text{otherwise} \quad f_{ij}^{q+1} = f_{ij}^{q} + \frac{g_{\vartheta k} - g_{\vartheta k}^{q}}{N_{\vartheta k} - N^{0}_{\vartheta k} - N^{M}_{\vartheta k} - N^{S}_{\vartheta k}}$$

There are, in conclusion, at least two different types of memory matrices, and one can use them either separately or together. This makes it important to distinguish between the memory matrix method and the particular algorithms which are based on it, and in order to underline such a distinction it is convenient to adopt a new terminology. The family of all algorithms which use memory matrices is referred to as MRM (*Memory Reconstruction Method*), whereas any individual member of this family is indicated with the label MRM followed by a number. More precisely, MRM-1 is the algorithm which employs only the vortex memory (density modulation), MRM-2 uses only the stationary points memory, and MRM-3 is the algorithm of equation 3.12 which exploits both memories.

At this point we are left with the problem of discovering yet more memory matrices, and here we have plenty of suggestions. It is plausible, for example, that a memory of *boundaries*, or more generally a memory of *discontinuities*, could be built, but we can leave these developments to the future. We have seen that the memory matrix method can indeed perform reconstructions from incomplete information, and therefore we already have what we were looking for: a model that may help us understand the *logic* of embryonic development.

The two general principles of MRM

When we speak of mathematical models for biology, we usually refer to formulae (such as the Hardy–Weinberg theorem, or the Lotka–Volterra equations) that effectively describe some features of living systems. In our case, embryonic development is not described by integrals and deconvolutions, and the formulae of the reconstruction algorithms cannot be a direct description of what happens in embryos. There is however another type of mathematical model. The formulae of energy, entropy and information, for example, apply to all natural processes, irrespective of their mechanisms, and at this more general level there could indeed be a link between reconstruction methods and embryonic development. For our purposes, in fact, what really matters are not the formulae *per se*, but the general conclusions that they allow us to reach, and among these there are at least two which are indeed worthy of attention.

In the MRM model, the initial memory matrix is a *tabula rasa*, a white page that is gradually filled during the reconstruction process, while the reconstructed picture starts with a uniform image, and becomes progressively differentiated in the course of time. A reconstruction with the MRM model, in other words, is a set of *two* distinct reconstructions that are performed in parallel. The point is that this *double* reconstruction is necessary for reasons that are absolutely general.

A picture and its projections are both structures of the real space, and, when projections are incomplete, there is no possibility of perfoming exact reconstructions if information comes only from structures of the real space (or from equivalent structures of the Fourier space). Only in a *related but autonomous* space we can find genuinely new information, and the memory space is precisely that type of independent world. It is in fact the only space where a system can get the extra information that allows it to increase its own complexity. The MRM model, in other words, leads to a universal concept: to the principle that *there cannot be a convergent increase of complexity without memory.*

The second fundamental characteristic of the MRM model is that information can be transferred from memory space to real space only by suitable *conventions*. In order to decrease the number of the unknowns in real space, it is necessary to give a *meaning* to the structures that appear in memory space, and this too is a conclusion whose validity is absolutely general. Real space and memory space must be autonomous worlds, because if they were equivalent (like real space and Fourier space, for example) they would convey the same information and no increase in complexity would be possible. But between two independent worlds there is no necessary link, and no information can be transferred automatically from one to the other. The only bridge that can establish a link between such worlds is an *ad hoc* process, i.e. a convention or a code. This amounts to a second universal principle: *there cannot be a convergent increase of complexity without codes.*

The Memory Reconstruction Method, in conclusion, gives us two general principles that must be valid for *all* systems which increase their own complexity, and embryos *are* such systems. The MRM model predicts therefore the existence of biological structures which are equivalent to reconstruction codes and to memory matrices. More precisely, the model leads to the conclusion that in embryos there must be codes and memories which are made of organic molecules, i.e. *organic codes* and *organic memories*. At this point, therefore, we can go back to biology and look for the existence of such structures in real life.

4

ORGANIC CODES AND ORGANIC MEMORIES

The existence of organic codes and organic memories is essentially an experimental problem, but experiments are planned and interpreted with criteria that are expressed in words. This inevitably requires some agreement about the terminology, and so this chapter starts with a few definitions. A code is defined as *a correspondence between two independent worlds*, and this definition immediately suggests a useful operative criterion. It means that the existence of a real organic code is based on (and can be inferred from) the existence of organic molecules – called *adaptors* – that perform two independent recognition processes. In the genetic code the adaptors are the transfer RNAs, but it will be shown that adaptors also exist in splicing and in signal transduction, which means that there are at least other two organic codes in real life. An organic memory is defined as *a deposit of organic information*, and this immediately qualifies the genome as an organic memory. It will be shown however that the determination state of embryonic cells is also a deposit of information, and so there are at least two different kinds of organic memory in real life. We conclude that codes and memories, as predicted by the mathematical model of epigenesis, do exist in the organic world. A conclusion that sets the stage for the problem of the next chapters, i.e. for the role that organic codes played in the history of life.

The characteristics of codes

The term *codes*, or *conventions*, normally indicates the rules which are adopted by a human community, but it has also a wider meaning.

A code can be defined as a set of rules that establish *a correspondence between two independent worlds*. The Morse code, for example, connects certain combinations of dots and dashes with the letters of the alphabet. The highway code is a liaison between illustrated signals and driving behaviours. A language makes words stand for real objects of the physical world.

The extraordinary thing about codes is that a new physical quantity appears in them, since they require not only energy and information but also *meaning*. The information of the word *ape*, for example, is measured by the bits that are required to choose the letters "a", "p" and "e" in that order, and is the same in all languages that have a common alphabet. But in English *ape* means a "tailless simian primate" whereas in Italian it stands for a "honey-making insect", and in both languages it could have had any other meaning. Words do not, by themselves, have meanings. They are mere labels to which meanings are given in order to establish a correspondence between words and objects. Because of this, it is often said that meanings are *arbitrary*, but that is true only if they are taken individually. The words of a language may seem arbitrary if taken one by one, but together they form an integrated system and are therefore linked by community rules. Codes and meanings, in other words, are subject to collective, not individual, constraints. Codes have, in brief, three fundamental characteristics (Figure 4.1):

(1) They are rules of correspondence between two independent worlds.

(2) They give meanings to informational structures.

(3) They are collective rules which do not depend on the individual features of their structures.

The independence between information and meaning implies, among other things, that their evolutionary changes are also independent. The word *pater*, for example, evolved into *padre, père, father, vater*, and in this case the order of the letters – i.e. the linear information of the word – changed, but its meaning did not. In the case of the word *ape*, instead, the meaning changed in both English and Italian, but the linear information did not. There is a qualitative difference between *informatic processes*, where only energy and

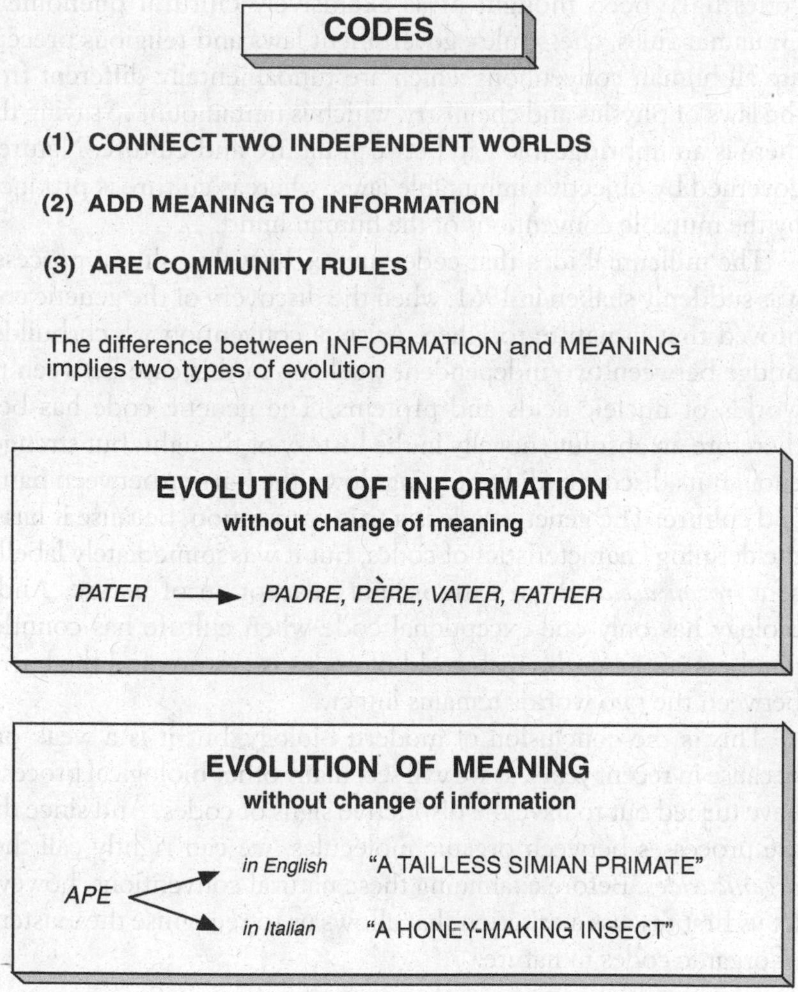

CODES

(1) CONNECT TWO INDEPENDENT WORLDS

(2) ADD MEANING TO INFORMATION

(3) ARE COMMUNITY RULES

The difference between INFORMATION and MEANING
implies two types of evolution

EVOLUTION OF INFORMATION
without change of meaning

PATER ⟶ PADRE, PÈRE, VATER, FATHER

EVOLUTION OF MEANING
without change of information

APE
    in English   "A TAILLESS SIMIAN PRIMATE"
    in Italian   "A HONEY-MAKING INSECT"

Figure 4.1   Codes, or conventions, are defined by general properties that are independent of the material composition of their objects, and are equally valid in the world of culture as in the world of organic life.

information are involved, and *semantic processes*, where rules appear which add meaning to information. Throughout recorded history, codes have been thought of as exclusively cultural phenomena. Grammar rules, chess rules, government laws and religious precepts are all human conventions which are fundamentally different from the laws of physics and chemistry, which is tantamount to saying that there is an unbridgeable gap between nature and culture. Nature is governed by objective immutable laws, whereas culture is produced by the mutable conventions of the human mind.

The millennial idea that codes are exclusively cultural processes was suddenly shaken in 1961, when the discovery of the genetic code proved that in nature too there exists a convention which builds a bridge between two independent worlds – in this case between the worlds of nucleic acids and proteins. The genetic code has been therefore an absolute novelty in the history of thought, but strangely enough its discovery did not bring down the barrier between nature and culture. The genetic code is a true convention, because it has all the defining characteristics of codes, but it was immediately labelled as a *frozen accident*, an extraordinary exception of nature. And if biology has only one exceptional code when culture has countless numbers of them, the real world of codes is culture and the barrier between the two worlds remains intact.

This is the conclusion of modern biology, but it is a weak one, because in recent years, as we will see, many other biological processes have turned out to have the distinctive signs of codes. And since they are processes between organic molecules, we can rightly call them *organic codes*. Before examining these natural conventions, however, let us first discuss a criterion that allows us to recognise the existence of organic codes in nature.

The organic codes' fingerprints

The genetic code is the only organic code which is officially recognised in the textbooks of modern biology, but it is also a model where we find characteristics that must belong to all organic codes. To start with, we can clearly appreciate the difference that exists, at the molecular level, between informatic and semantic processes. In biology, the seminal examples of these processes are, respectively, DNA transcription and RNA translation (Figure 4.2).

In transcription, an RNA chain is assembled from the linear information of a DNA chain, and for such assembly a normal biological catalyst (an RNA polymerase) is sufficient, because each step requires a single recognition process (a DNA–RNA coupling). In translation, instead, two independent recognition processes must be performed at each step, and to this purpose the catalyst of the reaction (the ribosome) needs special molecules that Francis Crick

CATALYSED ASSEMBLY (Transcription)

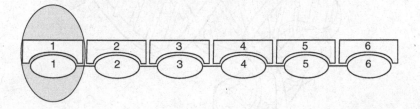

CODIFIED ASSEMBLY (Translation)

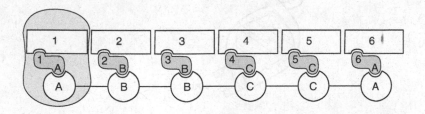

Figure 4.2   Transcription and translation are the prototype examples of catalysed and codified assemblies, i.e. of reactions that require respectively one and two recognition processes at each step of the assembly.

(1957) called *adaptors*, and that today are known as *transfer RNAs* (Figures 4.2 and 4.3).

Briefly, an amino acid is attached to a tRNA by an enzyme (an aminoacyltransferase) which specifically recognises a region of the tRNA, while a different region (the anticodon) interacts with a messenger RNA. With appropriate mutations, in fact, it has been possible to change independently the anticodon and the amino acid regions, thus changing the code's correspondence rules. These code changes have been artificially produced in the laboratory, but have also been found in nature. Mitochondria and some micro-organisms,

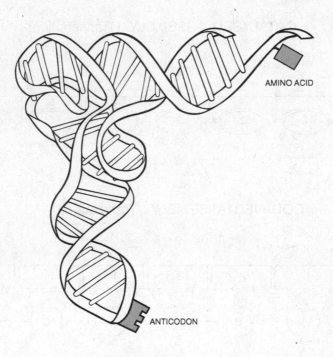

Figure 4.3   Transfer RNA is the prototype *adaptor*, i.e. a molecule that performs two independent recognition steps, one in the nucleotide world and the other in the world of amino acids.

for example, do have codes which differ from the universal one, which again shows that there is no necessary link between anticodons and amino acids. The codon recognition site is thus independent from the amino acid recognition site, and it is this independence that makes a code absolutely essential. Without a code, a codon could be associated with different amino acids and *biological specificity* – the most precious of life's properties – would be lost.

The function of an organic code, in conclusion, is to give specificity to a liaison between two organic worlds, and this necessarily requires molecular structures – the *adaptors* – that perform two independent recognition processes. In the case of the genetic code the adaptors are tRNAs, but any other correspondence between two independent molecular worlds needs a set of adaptors and a set of correspondence rules. The adaptors are required because the two worlds would no longer be independent if there was a necessary link between their units; and a code is required to guarantee the specificity of the link.

The adaptors were theoretically predicted by Francis Crick in order to explain the mechanics of protein synthesis, but they are necessary structures in all organic codes. They are the molecular *fingerprints* of the organic codes, and their presence in a biological process is a sure sign that that process is based on a code.

The bridge between genes and organisms

Between the genes and the final structure of a body there are, in all multicellular organisms, several levels of organisation whose general features have been known for some time (Figure 4.4). Genes intervene only at the very beginning of the ladder, when they are copied into primary transcripts of RNA. From this point onward, all other body construction steps take place in the absence of genes, and are collectively known as *epigenetic processes.*

(1) The first epigenetic event is the processing of primary transcripts into messenger RNAs. (2) The second is protein synthesis, the translation of messenger RNAs into linear chains of amino acids. (3) The third is protein folding, the transformation of linear polypeptides into

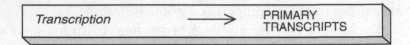

THE BRIDGE BETWEEN GENES AND ORGANISM

(A) THE GENETIC STEP

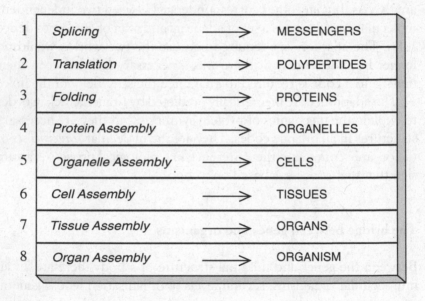

| *Transcription* | $\longrightarrow$ | PRIMARY TRANSCRIPTS |

(B) THE EPIGENETIC PROCESSES

1	*Splicing*	$\longrightarrow$	MESSENGERS
2	*Translation*	$\longrightarrow$	POLYPEPTIDES
3	*Folding*	$\longrightarrow$	PROTEINS
4	*Protein Assembly*	$\longrightarrow$	ORGANELLES
5	*Organelle Assembly*	$\longrightarrow$	CELLS
6	*Cell Assembly*	$\longrightarrow$	TISSUES
7	*Tissue Assembly*	$\longrightarrow$	ORGANS
8	*Organ Assembly*	$\longrightarrow$	ORGANISM

Figure 4.4   Between genes and proteins there are at least eight levels of epigenetic processes that together form what is collectively know as *epigenesis*.

three-dimensional proteins. Some proteins are then assembled into organelles (4), these are assembled into cells (5), and cells aggregate to produce tissues (6), organs (7) and finally the whole organism (8).

The units of each level of organisation are assembled into structures that become the units of the next level, thus giving rise to a hierarchy of assemblies, from proteins and organelles to cells, tissues and organs. *All events of epigenesis, in short, are processes of assembly.*

In living systems, assemblies usually require the presence of guiding factors that can be broadly called catalysts, and we can speak therefore of *catalysed assemblies*, but in some cases catalysts are not enough. The classic example is protein synthesis, the second epigenetic step of the *scala naturae* that builds a bridge between genes and organism. Protein synthesis is definitely an assembly operation, because amino acids are assembled into polypeptides, but a catalyst is not enough to determine the order of the units, and what is needed is a code-based set of adaptors. We must distinguish therefore between two very different types of epigenetic processes:

(1) processes of *catalysed assembly*, and

(2) processes of *codified assembly.*

The bridge between genes and organism, in conclusion, is realised by one genetic step and by at least eight types of epigenetic processes. Today it is generally believed that only protein synthesis is a codified assembly, but this has never been proved. On the contrary, we will see that many other epigenetic phenomena have the characteristic signs of true codified assemblies.

The splicing codes

The first step of epigenesis transforms primary transcripts into messenger RNAs by removing some RNA strings (called *introns*) and by joining together the remaining pieces (the *exons*). This is a true assembly, because exons are assembled into messengers, and we need therefore to find out if it is a *catalysed* assembly (like transcription) or a *codified* assembly (like translation). In the first case the cutting-and-pasting operations, collectively known as *splicing*, would require only a catalyst (comparable to RNA-polymerase), whereas in the second case they would need a catalyst and a set of adaptors (comparable to ribosome and tRNAs).

This suggests immediately that splicing is a codified process because it is implemented by structures that are very similar to those of protein synthesis. The splicing catalysts, known as *spliceosomes*, are huge molecular machines with molecular weights in the range of ribosome figures, and employ small molecular structures, known as *snRNAs* or *snurps*, which are comparable to tRNAs (Figure 4.5) (Maniatis and Reed, 1987; Steitz, 1988). The key point, however, is that the comparison goes much deeper than a mere similarity, because snRNAs, like tRNAs, have properties that fully qualify them as adaptors. They bring together, in a single molecule, two independent recognition processes, one for the beginning and one for the end of an intron,

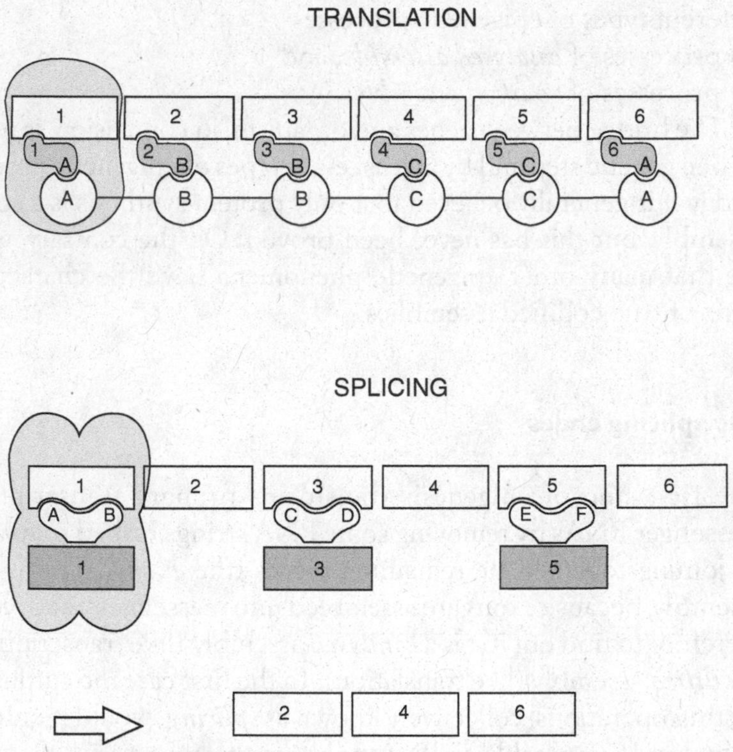

Figure 4.5   A schematic illustration of translation and splicing shows that both assemblies require adaptors, i.e. molecules that perform two independent recognition processes at each step of the reactions.

thus creating a specific correspondence between the world of transcripts and the world of messengers (Figure 4.6).

The two recognition steps are independent because the first step

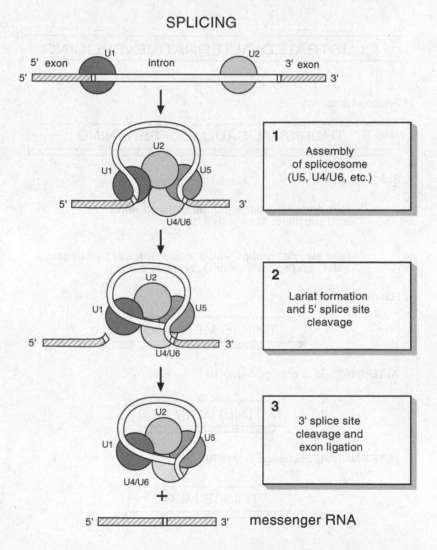

Figure 4.6   Splicing creates a correspondence between the world of primary transcripts and the world of messenger RNAs with molecules that independently define the beginning and the end of the spliced regions.

can be associated with different types of the second one, as demonstrated by the cases of *alternative splicing* (Figure 4.7). The choice of the beginning and of the end of an intron, furthermore, is

ILLUSTRATED ALTERNATIVE SPLICING

Primary transcript:

THORISMALISAULMEOFFERANING

Splicing rules:

(1) Take away all 3-letter groups that end with L
    and substitute them with a blank

(2) Take away all groups with 5 letters or fewer that begin
    with O and end with R or G, and join the rest

Resulting message:

THIS IS MEANING

ALTERNATIVE 1: change 5 into 10

THIS IS ME

ALTERNATIVE 2: change G into N

THIS IS MEG

Figure 4.7   An illustration of alternative splicing, with letters in place of nucleotides, underlines a deep parallel between the conventions of language and those of organic life.

the operation that actually defines the introns and gives them a meaning. Without a complete set of such operations, primary transcripts could be transformed arbitrarily into messengers, and again biological specificity would be lost. In RNA splicing, in conclusion, we find the three basic characteristics of the codes:

(1) Splicing establishes a correspondence between two independent worlds.

(2) Splicing is implemented by molecular adaptors which give meanings to RNA sequences.

(3) Splicing consists of a community of processes that guarantee biological specificity.

We must add however that there is an important difference between splicing and protein synthesis. Whilst the genetic code is practically universal, in the case of splicing there can be different sets of rules in different organisms and we are therefore in the presence of a plurality of splicing codes.

The signal transduction codes

Cells react to a wide variety of physical and chemical stimuli from the environment, and in general their reaction consists in the expression of specific genes. We need therefore to understand how the outside signals influence the genes, but for a long time all that could be said was that there must be a physical contact between them. The turning point, in this field, came with the discovery that external signals never reach the genes (Figure 4.8). They are invariably transformed into a different world of internal signals, called *second messengers*, and only these, or their derivatives, reach the genes. In most cases, the molecules of the external signals (known as *first messengers*) do not even enter the cell and are captured by specific receptors on the cell membrane, but even those which do enter (some hormones) must interact with intracellular receptors in order to act on genes (Sutherland, 1972).

The transfer of information from environment to genes takes place therefore in two distinct steps: one from first to second messengers, which is called *signal transduction*, and a second pathway from second

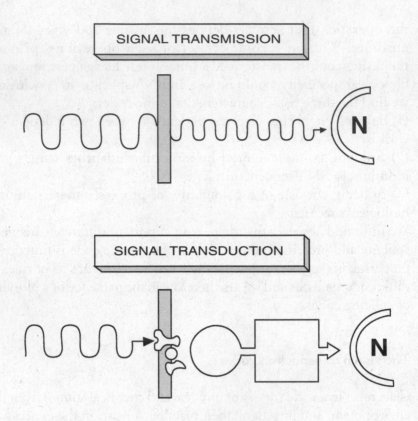

Figure 4.8   The signals that a cell receives from the environment do not reach the genes by transmission, but via a process of transduction which physically replaces them with different signals.

messengers to genes which is known as *signal integration*.

The study of signal transduction turned out to be a veritable mine of surprises. There are literally hundreds of first messengers (hormones, growth factors, neurotransmitters, etc.) whereas the known second messengers are only four (cyclic AMP, calcium ions, inositol trisphosphate and diacylglycerol), as shown in Figure 4.9 (Berridge, 1985, 1993). First and second messengers belong therefore to two independent worlds, which suggests immediately

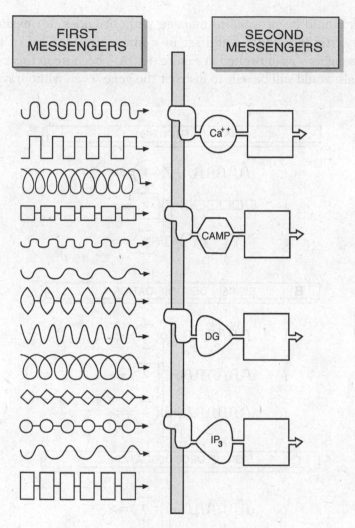

Figure 4.9   The first messengers of cell signalling are qualitatively and quantitatively different from the second messengers, and the latter consist of only four basic types of molecular cascades.

that signal transduction is likely to require the intervention of organic codes. But let us see if we can explain the experimental data in a different way.

It would be possible to manage without codes, for example, if every first messenger could set in motion a unique set of second messengers – as illustrated in Figure 4.10A – because in this case the signals would still be able to instruct the genes even without reaching

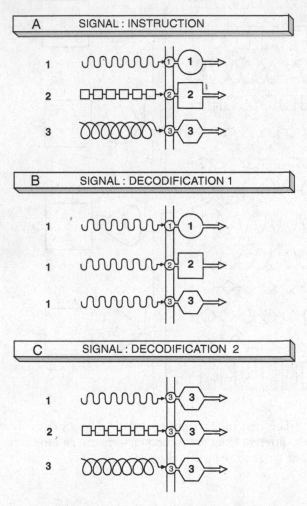

Figure 4.10   The first messengers of cell signalling can be processed in a variety of ways. Different signals can have the same effect and equal signals can lead to different results, thus showing that signal transduction is based on organic codes.

them. The facts, however, are very different. Acetylcholine, for example, is the signal that nerves deliver to most muscles, but does not have a unique meaning. Skeletal muscle cells respond by contracting, while cardiac muscle cells relax, and other cells remain indifferent (Figure 4.10B). And this pattern is not the exception but the rule. Environmental signals can be *decoded* in many different ways, and what reaches the genes is only the final result of an extremely complex decoding procedure. Another confirmation of this conclusion comes from the discovery that different signals can produce equal results. The liberation of calcium ions, for example, is a classic example of second messengers that are set free by a wide variety of external signals (Figure 4.10C).

The experimental results, in brief, have proved beyond doubt that outside signals do not have instructive effects. Cells use them to *interpret* the world, not to yield to it. Such a conclusion amounts to saying that signal transduction is based on organic codes, and this is in fact the only plausible explanation of the data, but of course we would also like a direct proof. As we have seen, the signature of an organic code is the presence of adaptors, and the molecules of signal transduction have indeed the typical characteristics of adaptors.

The transduction system consists of at least three types of molecules (Figure 4.11): a *receptor* for the first messengers, a *mediator*, and an *amplifier* for the second messengers (Berridge, 1985). The system performs therefore two independent recognition processes, one for the first and the other for the second messenger, and the two steps are connected by the bridge of the mediator. The connection however could be implemented in countless different ways since any first messenger can be coupled with any second messenger, which makes it imperative to use a code in order to guarantee biological specificity.

In signal transduction, in short, we find the three characteristics of the organic codes:
(1) A correspondence between two independent worlds.
(2) A system of adaptors which give meanings to molecular structures.
(3) A collective set of rules which guarantee biological specificity.
The effects that external signals have on cells, in conclusion, do not

SIGNAL TRANSDUCTION

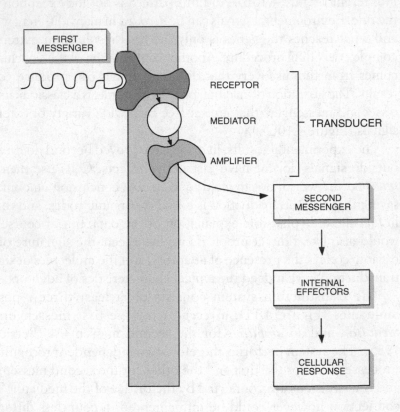

Figure 4.11   Signal transduction is implemented by a system of three molecules (a receptor, a mediator and an amplifier) that collectively function as the adaptor of an organic code.

depend on the energy and the information that they carry, but only on the meanings that cells give them with rules that can be called *signal transduction codes*.

Contextual information

We have seen that there are only four types of second messengers, and yet the signals that they set in motion do have specific effects, i.e. they are able to find individual genes among tens of thousands. How this is achieved is still a mystery, but some progress has been made and so far the most illuminating discovery in the field has been the demonstration that signalling molecules have in general more than one function. Epidermal growth factor, for example, stimulates fibroblasts and keratinocytes to proliferate, but has an anti-proliferative effect on hair follicle cells, whereas in the intestine it is a suppressor of gastric acid secretion. Other findings have proved that *all* growth factors can have three different functions, with proliferative, anti-proliferative, and proliferation-independent effects. They are, in short, *multifunctional molecules* (Sporn and Roberts, 1988).

In addition to growth factors, it has been shown that countless other molecules have multiple functions. Cholecystokinin, for example, is a peptide that acts as a hormone in the intestine, where it increases the bile flow during digestion, whereas in the nervous system it behaves as a neurotransmitter. Encephalins are sedatives in the brain, but in the digestive system are hormones that control the mechanical movements of food. Insulin is universally known for lowering the sugar levels in the blood, but it also controls fat metabolism and in other less known ways it affects almost every cell of the body.

The discovery of multifunctional molecules means that the function of many molecules is not decided by their structure, but by the environment of the target cell, i.e. by the *context* in which they find themselves. This implies that there is in cells an equivalent of the *contextual information* that plays such a relevant role in language (Harold, 1986). In everyday life, a message does not carry only grammatical information, but also information of a different kind which is called *implicit*, or *contextual*, because it is determined by the context of the message. The statement *"Come here"*, for example, has a unique grammatical structure, but the effect of the message changes tremendously with the person that delivers it (a friend or an enemy, a doctor or a policeman, a man or a woman) and with the circumstances of the delivery.

The existence of multifunctional molecules implies that a similar distinction exists in any cell between the syntactic information of a protein, which is uniquely determined by the order of the amino acids, and its contextual information, which is the actual function that the protein happens to be given by the cell. We have therefore the problem of understanding what it is that makes up the contextual information of a cell. This problem has not yet been solved, but a giant step forward was made with the discovery that second messengers do not act independently. Calcium ions and cyclic AMPs, for example, have effects which reinforce each other in some occasions whilst in others they are mutually exclusive (Rasmussen, 1989; Berridge, 1993). The cell, in short, does not merely transmit signals but is a system which manipulates and integrates them in many different ways. And it is precisely this ability that explains why a limited number of second messengers can generate an extraordinary number of outcomes, and end up with specific effects on genes.

The information carried by first messengers undergoes therefore two great transformations on its journey towards the genes. First, it is transformed into a world of internal signals with the rules of the transduction codes, and then these signals are given contextual information by being channelled along complex three-dimensional circuits. The actual construction of these integration circuits can only take place during embryonic development, and could be achieved either by catalysed or by codified processes. It is possible, therefore, that in addition to *signal transduction codes*, there are also *signal integration codes* in animal cells.

Determination and cell memory

At the beginning of the twentieth century, Hans Spemann made one of the most important discoveries of embryology. He was able to prove that the differentiation of developing cells – i.e. the actual expression of specific proteins – is always preceded by a process that determines the fate of those cells. Spemann made the discovery by studying what happens when small pieces of tissue are transplanted from one part of an embryo to another. He found that embryonic cells can change their histological fate (skin cells, for example, can become nerve cells) if they are transplanted *before* a critical period, but are totally unable to do so if the transplant takes place *after* that period. There is therefore, for every cell type, a crucial period of development in which *something* happens that decides what the cell's destiny is going to be, and that something was called *determination*.

Other experiments proved that determination does not normally take place in a single step but in stages, and that the number and duration of these stages vary from one tissue to another. The process of differentiation, in brief, is always preceded by the discrete changes of determination, and it is this phenomenon that we must understand if we want to explain the emergence of the various tissues of the body during embryonic development.

The most impressive property of determination is the extraordinary stability of its consequences. The process takes only a few hours to complete but leaves permanent effects in every generation of daughter cells for years to come. The state of determination, furthermore, is conserved even when cells are grown *in vitro* and perform many division cycles outside the body. When brought back *in vivo*, they express again the properties of the determination state as if they had never "forgotten" that experience.

In *Molecular Biology of the Cell* (1989) Alberts and colleagues declared explicitly that determination represents the appearance of a true cell memory:

"The humblest bacterium can rapidly adjust its chemical activities in response to changes in the environment. But the cells of a higher animal are more sophisticated than that. Their behaviour is governed not only by their genome and their present environment but also by their history ... During embryonic development cells must not only become different, they must also 'remain' different ... The differences are maintained because the cells somehow remember the effects of those past influences and pass them on to their descendants ... Cell memory is crucial for both the development and the maintenance of complex patterns of specialization ... The most familiar evidence of cell memory

is seen in the persistence and stability of the differentiated states of cells in the adult body."

Despite the authorithy and the success of *Molecular Biology of the Cell*, the idea of cell memory has not become popular, even if it is *necessarily* linked to the concept of determination. A memory is a deposit of information, and we can give the name of *organic memory* to any set of organic structures that is capable of storing information in a permanent (or at least in a long-lasting) way. The genome, for example, is not only an hereditary system but also an organic memory, because its instructions are not only transmitted to the offsping, but are also used by the organism itself throughout its life. We can rightly say, therefore, that the genome is the *genetic memory* of a cell.

The state of determination has also the characteristics of an organic memory, because it has permanent effects on cell behaviour, but it is an *epigenetic* memory, i.e. a memory which is built in stages during embryonic development by epigenetic processes. We conclude therefore that embryonic cells have two distinct organic memories: the *genetic memory* of the genome and the *epigenetic cell memory* of determination.

The other face of pattern

Genetics is based on the concept that the structures of an organism are controlled – more or less indirectly – by its genes, and since embryos are no exception, there must be genes even for their developing structures. For more than half a century, however, classical genetics was unable to prove their existence, because genes are recognised by mutations, and mutations of developmental genes normally bring development to a halt, thus making it impossible to observe the effect that they have *in vivo*. This obstacle was overcome only thanks to some peculiar characteristics of *Drosophila melanogaster* (a fruit fly), and to the patient work of Edward Lewis at the California Institute of Technology.

The *Drosophila* characteristics that allowed the discovery of the first developmental genes are the so-called *homeotic mutations* which

William Bateson first described in 1895. These mutations do not arrest development but transform one part of the body into another. *Antennapedia* mutations, for example, transform antennae into legs, which gives rise to an insect with two legs sprouting from its head, whereas *bithorax* mutations transform the third thorax into a second one, giving the insect an extra pair of wings.

In order to identify the homeotic genes, it was necessary to distinguish them from the other developmental genes, but the mutations of these were usually lethal and prevented the animals from going beyond the larval stage. Lewis noted, however, that *Drosophila* larvae have 12 segments which can be individually recognized under the microscope because of a belt of hairs and denticles that has a distinct morphology in every segment. Even if the insects were dying in the larval stage, therefore, the effects of developmental mutations could still be seen in the altered microscopic morphology of their segments. After thousands of crosses and of *post mortem* examinations of larval cadavers, Lewis was able to trace the loci of *bithorax* mutations to the right arm of chromosome 3. And soon afterwards the loci of *Antennapedia* mutations were also found on the same chomosome.

It must be underlined that homeotic mutations do not interfere with the normal processes of tissue differentiation, since a mutated fly has all the tissues and organs of a normal fly. The only difference is that some of the right structures end up in wrong places, which shows that homeotic genes control the geographic organisation of structures. Spatial organisation is often referred to as *pattern*, and we conclude therefore that homeotic genes are *pattern genes*.

It has long been accepted that there must be a spatial organisation – a pattern – in embryonic development. An arm and a leg, for example, or a right arm and a left arm, are made of the same tissues and yet their shapes are different, which means that during limb develoment the same cells can be arranged in different spatial configurations. The final form of a body structure, however, could be the automatic result of a chain of differentiation processes, exactly as the form of a crystal is the inevitable outcome of crystallisation reactions, and for a long time many embryologists have indeed thought that form does not need genetic control. Lewis made therefore two discoveries in one stroke, since he found not only the first genes of development but also the first genes of pattern.

Shortly afterwards came the demostration that homeotic genes belong to a wider class of pattern genes that control the entire body plan of the organism (the *Bauplan*) during embryonic development. Sander, Nüsslein-Volhard, Wieschaus, Garcia-Bellido and others were able to prove that the spatial organisation of *Drosophila* embryos is built up in stages under the control of three main groups of genes: (1) polarity (or axis) genes, (2) segment genes, and (3) homeotic genes (now known as *HOM* genes). Then came the discovery that similar genes also exist in vertebrates. They are now called *Hox* genes because of their *homeobox*, a sequence of 183 nucleotides which they have in common with the homeotic genes of invetebrates, and it has been shown that *Hox* genes too are involved in spatial organisation. Pattern genes, in brief, exist in all animals, and are used in every embryo in the construction of the body plan.

The great attention which has been given to the *spatial* characteristics of pattern appears, however, to have overshadowed its *temporal* properties. Pattern genes are expressed in a very early period of development, but their effects last for the whole life of an organism. Pattern genes, in other words, are genes that contribute to *cell memory*. They are expressed very early, and only for a very short period of time, but pattern characteristics survive indefinitely, and somehow, therefore, a cell keeps a "memory" of what they did. We could even say that, after the brief expression of its genes, pattern only lives in the memory space of a cell.

Pattern does not account for the whole of cell memory, because there are many other characteristics of determination that have long-lasting effects, but it is certainly an important part of that memory. The discovery of the pattern genes, in conclusion, can also be regarded as the discovery of the first genes that contribute to cell memory.

Hints from developing neurons

The embryonic development of the nervous system is one of the most interesting processes of ontogenesis, and its study has also been one of the richest sources of information on the strategies that embryos adopt to solve their problems (Bonner, 1988; Edelman, 1988).

The first phase of neural development begins when a strip of ectoderm is induced to become nervous tissue by the underlying mesoderm, and comes to an end when neuroblasts complete their last mitosis, an event which marks the "birth" of a neuron. The extraordinary thing is that all that a neuron will do in its whole life is determined by its birth, and more precisely by the time and by the place of its birth. These two parameters leave an indelible mark in the cell, and somehow therefore they must physically be recorded in a true organic cell memory.

The second phase is a period during which neurons migrate to the places of their final destination, places that they "know" because somehow they are "written" in the memory of their time of birth and of their birthplace.

The third phase of neural development begins when neurons reach their definitive residence. From this time on, the body of a neuron does not move any more but sends out "tentacles" that begin an exploration journey in the surrounding space. Any tentacle (or *neurite*) ends with a roughly triangular lamina, called the *growth cone*, which moves like the hand of a blind man, touching and feeling any object on its path before deciding what to do next. The axons of motor neurons are the longest of such tentacles, and their first task is to leave the neural tube for the rest of the body, in search of organs that require nerve connections.

If neurons were to depend on precise geographical information about their paths and their targets, they would need far bigger genomes than they could possibly carry. It is clear therefore that they rely on a different type of information, but it is also clear that the final result is the same, since all organs end up with nerve connections in precise geographical areas. With a small amount of a *special* information, neurons manage to obtain the same result that could be achieved with an enormous amount of *geographic* information. How do they do it?

The strategy is in two stages. In the first part of their journey, growth cones move along tracks provided by a few types of molecules, with a preference for those of other axons (which explains why growth cones migrate together and why their axons form the thick bundles that we call nerves). The lack of a precise geographical knowledge of the targets is compensated by an overproduction of nerve cells, which ensures that some of these will actually reach the targets. With such a mechanism, however, it is inevitable that many growth cones end up in wrong places, and it is at this point that the second part of the strategy comes into play.

The organs that are to be innervated send off particular types of molecules, known as *nerve growth factors*, whose effect is truly dramatic: the neurons which get them survive, while the others die. More precisely, neurons are programmed to commit suicide – i.e. to activate the genes for programmed cell death, or *apoptosis* – at the end of a predetermined period of time, and nerve growth factors are the only molecules that can de-activate their self-destruction mechanism. The result is that neurons which reach the right places survive, and the others disappear.

There are many aspects of this phenomenon that are worth discussing, but here let us concentrate on just one point. It is known that any cell can switch on its suicide genes in response to external signals, but it is also known that signalling molecules can have many different functions (all growth factors, as we have seen, are multifunctional molecules). This means that the recognition of a signalling molecule and the activation of suicide genes are two independent processes, which gives us the problem of understanding how cells manage to connect them. One possibility is that cells have genetic instructions for any possible environmental situation, but this is not a realistic hypothesis because the genome would have to be enormous. Another possibility is that cells use *apoptosis codes*, i.e. a limited number of rules that give an apoptotic meaning to signalling molecules. The existence of organic codes for programmed cell death has never been suggested before, but without them it is practically impossible to explain the facts.

A similar conclusion is obtained for the mechanism of cell migration. Cells must express specific genes in order to move, and they do so in response to signalling molecules. In this case too, however, the link between signals and genes can be made in countles different ways, and the only realistic explanation is that the mechanism is based on a limited number of *cell migration codes.* Both apoptosis and cell migration, moreover, depend on the determination state of the cell, and this brings us to the conclusion that there must be a link between organic codes and cell memory.

Neural embryonic development, in conclusion, appears to be understandable only if we admit that its mechanism is based on organic codes and organic memories. Without these structures, an explanation becomes virtually impossible, and one can only hope that future research will be able to prove conclusively the reality of their existence.

The key structures of embryonic development

Embryologists have always maintained (and most of them still do) that embryology cannot be reduced to genetics, but often their arguments have not been totally convincing. The claim that development comprises both genetics and epigenesis, and not genetics alone, is a valid one, but is not enough to prove the point.

Some epigenetic processes (for example protein synthesis and ribosome self-assembly) also take place in prokaryotes, and yet these cells do not give rise to embryos. It is not epigenesis as such, therefore, that accounts for development, but a particular type of epigenesis that prokaryotes do not have. It could be pointed out that prokaryotes lack the complex structures of the eukaryotic cell, but this does not explain their lack of embryonic potential. Protozoa, for example, do have the eukaryotic cell structure but they too are incapable of producing embryos.

It is not epigenesis as such, nor eukaryotic cell structure, that accounts for embryonic development, and we are bound to conclude that development is based on "something" which does not exist in unicellular organisms, and which belongs to the family of the epigenetic structures because it is produced in stages. The only structure which has those requirements is cell memory, and finally this does give us an answer.

Once again, however, we are not completely comfortable. The conclusion that embryonic development is based on cell memory because that is the only structure which has the right requirements, is a conclusion obtained by default, and that is not satisfactory. We want to know *why* things happen, and a solution obtained by elimination does not give us that insight. We want to know, for example, if cell memory is really essential to development, and, if it is, we want to know why. This kind of question can be answered only by very general arguments, possibly with the assistance of mathematical algorithms, and it is at this point that the Memory Reconstruction Method can help us.

The reconstruction of a three-dimensional structure (a cell) from incomplete linear information (a set of genes) can be achieved only by reconstructing in parallel two distinct but mutually dependent three-dimensional structures (a cell and a cell memory). The Memory Reconstruction Method gives us an abstract model of an embryonic cell because it gives us a model of a structure that contains an epigenetic memory, but the crucial point is that the model makes us understand why that memory is essential. Without such a memory it would not be possible to obtain *a convergent increase of complexity*, and the real logic of embryonic development is precisely that kind of increase. Now, at last, we have a proper answer. Cell memory is a key structure of embryonic development because it is essential to the convergent increase of complexity that is typical of development.

As always happens, a new conclusion inevitably raises new questions, and in this case the first problem that calls for our attention is the origin of cell memory, a problem which takes us back to evolution and to the evolutionary mechanisms. Now, however, we have something new to discover in the past, and we can look at the history of life from a point of view which has not been considered before.

5

THE ORIGIN OF LIFE

New theories on the origin of life appear at almost regular intervals in the scientific literature, thus creating the impression that there are virtually endless solutions to the problem. In reality most theories are but variations on two basic themes: the *metabolism-first* paradigm proposed by Oparin and the *replication-first* paradigm of J.B.S. Haldane. This is because every organism is conceived as a duality of genotype and phenotype (software and hardware), and since a sudden appearance of such a system would be little short of a miracle, there seemed to be no choice but to start either from primordial genes or from primordial proteins. In 1981, however, I proposed that an alternative does exist, because the cell is not a duality but a trinity of genotype, phenotype and ribotype. And the ribotype has a logical and a historical priority over genotype and phenotype. At the end of the chapter this proposal is illustrated with a metaphor that compares the cell to a city, where proteins are the houses and genes are their blueprints. The traditional paradigms would ask *"Which came first, the houses or the blueprints?"* while the ribotype theory amounts to saying that *"It was the inhabitants who came first. It was they who made houses and blueprints."* The aim of this chapter, in short, is to present an overview of *all* paradigms on the origin of life: the *gene-first*, the *protein-first* and the *ribotype-first* approach. Another goal is to underline that the very beginning of macroevolution (the origin of life) was associated with the appearance of the first organic code on Earth (the genetic code).

The primitive Earth

The age of meteorites tells us that the solar system – and therefore the Earth – was born roughly 4.6 billion years ago. The oldest terrestrial rocks are zircone crystals (zirconium silicates) which are 4.2 billion years old, but these stones do not tell us much apart from their age, because they are igneous, or magmatic, rocks whose melting processes have erased any trace of history. Much more interesting are the sedimentary rocks, because these were formed by materials that sank to the bottom of ancient seas, and may still contain remnants of the past. The oldest sediments have been found at Isua, in Greenland, and are 3.8 billion years old, which means that there were immense streches of water on our planet at that time, and that the first oceans had originated many millions of years earlier.

But what was there, in addition to water, on the primitive Earth? The four outer planets of the solar system (Jupiter, Saturn, Uranus and Neptune) are still made up mainly of hydrogen, helium, methane, ammonia and water, and it is likely that those same chemicals were abundant everywhere else in the solar system, and therefore even in its four inner planets (Mercury, Venus, Earth and Mars). These were too small to trap light chemicals, such as hydrogen and helium, but the Earth had a large enough mass to keep all the others. It is likely therefore that the Earth's first atmosphere had great amounts of methane (CH_4), ammonia (NH_3) and water, and was, as a result, heavy and reducing, like Jupiter's.

The Isua sedimentary beds, on the other hand, contain iron compounds that could have formed only in the *absence* of oxygen, which means that the Earth did not have an oxidizing atmosphere for hundreds of millions of years. Those same sediments, however, contain also many types of carbon compounds, and this shows that, by Isua's times, there were in the atmosphere substantial quantities of carbon dioxide (CO_2), and probably nitrogen (N_2), two gases which are neither reducing nor oxidizing.

It seems therefore that the primitive atmosphere did change considerably in the first billion years, and from a highly reducing state, dominated by methane and ammonia, slowly turned into a slightly reducing one, where the dominant gases were carbon dioxide and nitrogen. This change can be explained both by strong volcanic activity, which poured enormous amounts of carbon dioxide into the air, and by the fact that the very first atmosphere was probably stripped from the Earth by the solar wind that was produced by huge explosions on the Sun. The lack of free oxygen means that the primitive Earth did not have an ozone layer to protect it from the Sun's ultraviolet rays, and this radiation was therefore pouring freely onto the Earth's surface, adding yet more energy to the vast amounts that were already produced by lightning, by radioactivity, by hydrothermal vents, and by the prolonged meteorite bombardment that left such visible scars on the face of the Moon.

These data give us a fairly realistic account of the main chemicals, and of the major energy sources, of the primitive Earth, and are the obligatory starting-point for reconstructing the first steps of life's history. The seminal experiment which set in motion this research field was performed by Stanley Miller, in Chicago, in 1953. Miller built an apparatus where water was kept under an atmosphere of ammonia, methane and hydrogen, and simulated the primitive Earth's conditions by having the water boiling and by producing electric sparks in the atmosphere (Figure 5.1). In a few days, the water turned into a reddish solution which contained a variety of organic compounds, such as aldehydes, urea, formic acid and amino acids. The experiment was repeated many times, and with many other variants, for example with atmospheres dominated by nitrogen and carbon dioxide, and a spontaneous production of organic compounds was observed in all cases, provided that no free oxygen was present.

Of the many extraordinary results of these experiments, one of the most illuminating is the fact that, among the intermediate products, there were hydrogen cyanide (HCN) and formaldehyde (CH_2O). Hydrogen cyanide can lead to adenine, which can be regarded as a HCN polymer (Figure 5.2), whilst formaldehyde can be turned into ribose (Figure 5.3). Together with phosphates, furthermore, these two compounds can form ATP, the molecule that cells use as a universal energy source. Formaldehyde and hydrogen cyanide, moreover, allow the synthesis of many amino acids such as glycine (Figure 5.4).

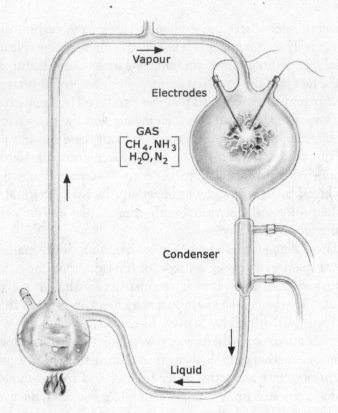

Figure 5.1   Stanley Miller's apparatus for the synthesis of organic compounds in abiotic conditions. Hydrogen cyanide, ammonia and amino acids are among the most interesting molecules which are obtained in this type of experiment.

Another result of great interest is that the four components of the RNAs (adenine, guanine, cytosine and uracil) have all been obtained in abiotic conditions, but thymine has never been found, which means that primitive nucleic acids could have been RNAs but not DNAs.

Even if the atmosphere of the primitive Earth could change within ample limits, therefore, we are bound to conclude that the first oceans became filled with organic molecules. In addition to molecular syntheses there also were, of course, degradation processes going on,

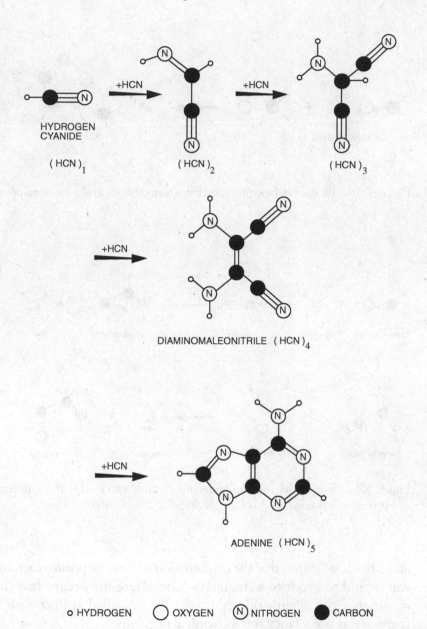

HYDROGEN
CYANIDE

(HCN)₁

(HCN)₂

(HCN)₃

DIAMINOMALEONITRILE (HCN)₄

ADENINE (HCN)₅

○ HYDROGEN   ○ OXYGEN   Ⓝ NITROGEN   ● CARBON

Figure 5.2   Adenine can be obtained from the polymerisation of hydrogen cyanide.

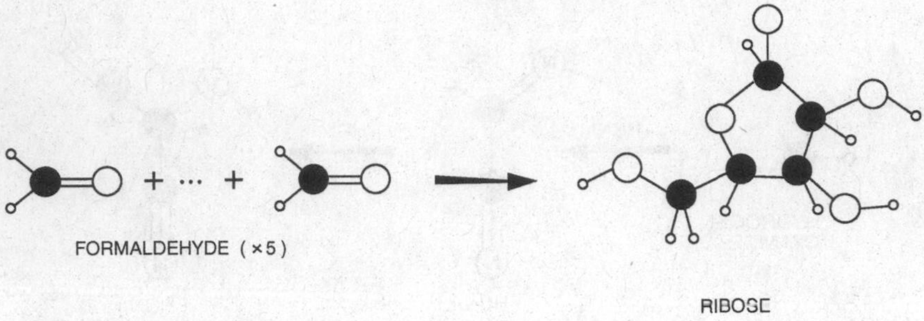

Figure 5.3   Ribose can be synthesised from formalheyde in abiotic conditions.

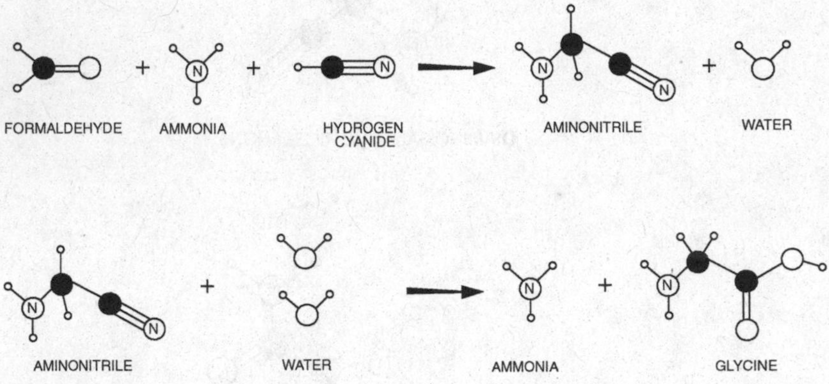

Figure 5.4   Formaldehyde and hydrogen cyanide can lead to the synthesis of various amino acids, in this case glycine.

and it has been shown that the combination of these opposite reactions was bound to produce a stationary state where the oceans had the consistency of a slightly diluted broth. The so-called *primitive soup*, therefore, is not a fancy but a chemical necessity.

Together with the positive results, however, we must also consider the negatives ones, and of these one of the most important is the fact that the abiotic synthesis of molecules has been relatively easy for amino acids, but much more difficult for nucleotides. Another complication is the fact that left-handed and right-handed molecules are produced *in vitro* with the same frequency, and this is a serious obstacle, because most biological reactions require only one type of symmetry. As we can see, the formation of a primitive soup was an important step, but was only a first step, and does not take us very far. In order to go further, we clearly need to discover something else.

Chemical evolution

In order to join two amino acids with a peptide bond, a water molecule has to be removed, and this suggests immediately that the reaction should not easily take place in water. The energy balance does confirm, in fact, that amino acid polymerisation is not favoured in water, and we cannot expect therefore that the primitive broth would spontaneously produce a stable population of proteins. How then did these molecules appear?

A solution to this problem was proposed in the 1960s by Sidney Fox, on the basis of experimental results that he obtained by heating up a mixture of amino acids in the absence of water. Fox found that, in these conditions, amino acids do aggregate into macromolecules which can even reach large dimensions, and which he called *proteinoids*. These are not real proteins because their amino acids are not arranged in linear chains of polypeptides, but form directly a variety of three-dimensional chemical bonds. Proteinoids, however, are somewhat similar to proteins in various respects, including a weak catalytic activity (they can, for example, catalyse ATP hydrolysis).

The important point is that hot and dry conditions were surely existing somewhere on the primitive Earth, for example in ponds and lakes that the Sun or volcanoes had dried up. It is reasonable, therefore, to expect that proteinoids did exist on our planet. But Fox went further, and proved that proteinoids can easily generate higher structures. If a concentrated proteinoid solution is heated to between 120 and 200 °C and then is very slowly cooled down, one can observe that proteinoids spontaneously form vesicles which Fox called *microspheres*. These structures come in fairly regular forms and dimensions (their diameters vary between 1 and 2 μm only), are very stable, and retain the weak catalytic activity of individual proteinoids. Despite the fact that lipids are absent, furthermore, a high number of vesicles exhibit semipermeable boundaries that look strangely similar to the lipid bilayer of true plasmatic membranes. But perhaps the most interesting thing is that microspheres can absorb proteinoids from the surrounding solution, which allows them to grow and eventually to divide in two by fission or budding.

Fox's microspheres, in conclusion, are the first systems obtained *in vitro* that present a rudimentary type of metabolism. The evolutionary potential of the microspheres, however, remains a mystery. It is fairly likely that they appeared on the primitive Earth, but we cannot be sure that they had a future, and it is for this reason that other solutions have been explored.

One of the most interesting is the theory of *surface metabolism*, an approach that was proposed, in different forms, by John Bernal in 1951, by Graham Cairns-Smith in 1982 and by Günter Wächtershäuser in 1998. The central idea of this theory is based on solid thermodynamic arguments. The formation of a peptide bond is not favoured in solution because it increases the entropy of the system, but on a surface the same process takes place with *a decrease of entropy*, and is therefore favoured. And this is true not only for peptide bonding but for many other types of polymerisation. A great number of enzymatic reactions require a collision of three molecules, an event which is highly unlikely in space but much more probable on a surface.

It is a thermodynamic principle, in short, that spontaneous reactions are more likely to occur on surfaces than in space, and it is reasonable therefore to conclude that the first metabolic structures were two-dimensional systems and not three-dimensional ones. Rather than in a primitive broth, in other words, chemical evolution could well have started on *primitive pizzas*.

Bernal and Cairns-Smith proposed that the first metabolic surfaces were provided by crystals of clay, and therefore that life did literally originate in mud, because clays can adsorb a vast range of organic molecules. Unfortunately clays can only favour an accumulation of pre-existing molecules, not the *in situ* develoment of two-dimensional organic systems, because their negative charges repel the negative charges of amino acids and nucleic acids. This is why Wächtershäuser proposed that surface metabolism developed on positively-charged minerals, and in particular on crystals of iron pyrite (FeS_2).

Wächtershäuser's theory suggests, furthermore, that the first two-dimensional organic systems could eventually detach themselves from their supporting surfaces, and gradually turn into three-dimensional vesicles. The end result is, again, the formation of primordial metabolic structures which are supposed to be *potentially capable* of evolving into primitive cells. But even in this case, we do not know if those structures were destined to abort or to go on evolving all the way up to the origin of cellular life.

Postchemical evolution

The first scientific theories on the origin of life were proposed by Alexander Oparin in 1924 and by J.B.S. Haldane in 1929. Oparin discovered that a solution of proteins can spontaneously produce microscopic aggregates – which he called *coacervates* – that are capable of a weak metabolism, and proposed that the first cells came into being by the evolution of primitive metabolic coacervates. Haldane, on the other hand, was highly impressed by the replication properties of viruses, and attributed the origin of life to the evolution of virus-like molecular replicators.

Today, Oparin's coacervates are not as favoured as Fox's microspheres or Wächtershäusers's vesicles, and RNA replicators are preferred to Haldane's viroids, but these differences have not changed the substance of the original opposition. Between the two fundamental functions of life – metabolism and replication – Oparin gave an evolutionary priority to metabolism, while Haldane gave it to replication, and the choice between these two alternatives is still the key point that divides the origin-of-life theories in two contrasting camps.

We can say therefore that the *metabolism-first* idea (the *metabolism paradigm*) goes back to Oparin, while the *replication-first* concept (the *replication paradigm*) goes back to Haldane. And since metabolism is based on proteins, and replication on nucleic acids, Oparin's paradigm is equivalent to saying that proteins (the hardware) came first, whereas Haldane's paradigm maintains that it was nucleic acids (the software) that had priority.

At this point, it becomes important to understand whether we really must choose between these alternatives, or whether there is a third option. Let us notice that the systems which came *immediately before the first cells* were necessarily made of both proteins and nucleic acids, and there must have been therefore a precellular period in which the two types of molecules *evolved together*. But in order to evolve together they had to coexist, and so we need to understand how that coexistence came into being. This problem admits three possible solutions:

(1) The first mixed systems of proteins and nucleic acids came suddenly into existence.

(2) The first mixed systems came from the evolution of one or more systems made of proteins where eventually nucleic acids also appeared (solution *metabolism first*).

(3) The first mixed systems came from the evolution of one or more systems made of nucleic acids where eventually proteins also appeared (solution *replication first*).

As for the first solution, it is not absurd to think that primitive proteins and nucleic acids could occasionally meet, but it would be unreasonable to pretend that such chance encounters could give rise to long-lasting, integrated, evolvable systems. Such a solution would be equivalent to admitting that life originated suddenly, in a single extraordinay quantum jump, and this is just too miracle-like to be taken into consideration. If we admit that the origin of life was the result of natural events which had a *realistic* probability of taking place, the only plausible solutions are to be found either in Oparin's or in Haldane's paradigm. More precisely, we must admit that these are the sole paradigms which offer an explanation for *the first phase of precellular evolution*, the phase that went from the first organic molecules to the first mixed systems made of proteins and nucleic acids. This however does not authorize us to say that the same mechanisms also operated in *the second phase of precellular evolution*, the phase that went from the first mixed organic systems to the first cells. We conclude that *precellular evolution must be divided into two major stages: one before and one after the appearance of the first integrated systems made of both proteins and nucleic acids.*

The first phase corresponds to classical *chemical* evolution, but the second one is more difficult to define, because it is no longer chemical evolution but not yet biological evolution. It is however necessary to characterise it, and to this purpose we can give it the name of *postchemical* evolution. Before the origin of life, in other words, there must have been two evolutionary stages that were temporally and conceptually distinct: one of *chemical* evolution and the other of *postchemical* evolution.

Such a distinction is important because it gives us a criterion for a better evaluation of the origin-of-life theories. The solutions proposed by Sidney Fox or Wächtershäuser, for example, are exclusively theories of chemical evolution, and tell us nothing about postchemical evolution. It would be wrong to criticise them for this, but it would also be wrong to say that, if they explain chemical evolution, they also explain postchemical evolution and therefore the origin of the cell.

The concept of postchemical evolution, in conclusion, allows us to realise that there is another important dichotomy in the origin of life field. In addition to the distinction between metabolism-first and replication-first theories, it is necessary to distinguish between theories of *chemical* evolution and theories of *postchemical* evolution.

The metabolism paradigm

The sudden appearance on Earth of a system capable of both metabolism and replication is too unlikely to be taken seriously. All reasonable theories on the origin of life assume therefore that chemical evolution started from systems that could perform only one of those functions. Hence the great schism between *metabolism-first* theories (Oparin's paradigm) and *replication-first* scenarios

(Haldane's paradigm).

In favour of the metabolism paradigm there are, first of all, the results of the simulation experiments, and in particular the fact that the abiotic production of amino acids is so much easier than that of nucleic acids. Chemistry tells us that the primitive Earth could indeed generate enormous amounts of organic molecules that were potentially capable of having some type of metabolism, and of producing structures as complex as Oparin's coacervates, Fox's microspheres, or Wächtershäuser's vesicles.

The problem, of course, is to evaluate the evolutionary potential of these structures. It is true that Fox's microspheres, for example, can grow and divide by budding or fission, but they lack any form of heredity, and the simulation experiments are too brief to inform us about their long-term potential. The only way of obtaining this kind of information is by using mathematical or chemical models, and such a solution, however imperfect, does have a certain degree of plausibility.

The first model of a system that is capable of growing by metabolism and of dividing by fission was proposed – with the name of *chemoton* – by Gánti (1975). Such a system receives metabolites from the environment, expels waste products, and performs a metabolic cycle that begins with one molecule and ends by making two of them. The system is therefore autocatalytic, but it is not using enzymes, and this leaves us in the dark about its biological potential.

In order to have a metabolic and enzymatic system, it would be necessary to have proteinaceous enzymes which can catalyse the synthesis of other enzymes, and for this they should be capable of making *peptide bonds*. Such systems have not been found in nature, so far, but according to Stuart Kauffman (1986) they *could* have existed in the past, and in primitive compartments could have produced autocatalytic networks which had the potential to "jump" from chaos to order. Even if we admit that those enzymes existed, however, we still have the problem of accounting for the origin of the complex autocatalytic networks that housed them.

An elegant solution to this problem has been proposed by Freeman Dyson (1985) with the model of a generalised metabolic system, whose behaviour is totally random and whose chemical composition is not specified in advance. In order to describe the evolution of such a system, Dyson used Kimura's equations of genetic drift, and found that, in certain conditions, a system of inert molecules does have a finite chance of jumping from a state of inertia to a state of metabolic activity. Dyson's model is interesting because it has at least three important implications:

(1) A primitive metabolic system had to have a certain initial complexity to start with: it cannot contain fewer than 10 000 monomers for its molecules, and the monomers must be of at least ten different types (which means that amino acids are in but nucleic acids are out).
(2) The system is very tolerant of errors, and can therefore survive and leave descendants even without mechanisms of exact replication.
(3) The system can tolerate, within very wide limits, the presence of molecules which are either inert or parasitic, and therefore do not contribute to metabolism.

This last property is particularly important, because it allows Dyson to make a hypothesis on the origin of nucleic acids. The assumption is that primitive metabolic systems learned to use ATP molecules as energy sources, thus transforming them into AMP molecules that were accumulated as waste products. These packed deposits, in turn, created the conditions for the polymerisation of nucleotides, thus leading to the origin of the first RNAs. At the beginning, the RNA molecules were useless and even potentially dangerous compounds, but the system could tolerate them, and eventually the RNAs became perfectly integrated into their hosts.

Up until the origin of RNA molecules, Dyson describes the logical consequences of the initial hypotheses, and his scheme is therefore a coherent theory of *chemical* evolution. But the mathematical model does not say anything about the subsequent integration of RNAs and hosts, and on this point Dyson resorts to a *supplementary* conjecture. He proposes that primitive RNAs invaded their metabolic hosts, and used them for their own replication, like viruses do, which is exactly Haldane's hypothesis. Dyson concludes therefore that, after Oparin's metabolism stage, came Haldane's replication stage, and his final scheme becomes: *"metabolism first, replication second"*. That RNAs could replicate themselves within precellular systems, as viruses do, is highly unlikely, but this point has nothing to do with Dyson's mathematical model, and can be regarded as an unnecessary addition. If we stick only to the intrinsic characteristics of Dyson's model, we have something very useful in our hands, because the scheme does give a valid answer to the main problem of chemical evolution: the problem of explaining how primitive systems made of proteins could be able to produce RNAs.

A somewhat parallel solution to the same problem has also been given by Wächtershäuser, with the description of an hypothetical, but plausible, sequence of chemical reactions that lead to the same final result. We have therefore both mathematical and chemical models that are capable, in principle, of explaining chemical evolution. This is only the first part of precellular evolution, and there still is a long way to go before the origin of the cell, but at least the metabolism paradigm does give us a good starting point.

The replication paradigm

The discovery of viruses made an enormous impression on biologists, because it proved that something much smaller than a cell maintained the ability to replicate, the most quintessential of life's properties. Haldane knew only too well that viruses are totally dependent on cells for their replication, and therefore that they could have evolved only after cells, but those tiny proliferating crystals in the interior of huge cellular structures appeared to state a deeper truth: *that replication is simpler than metabolism*. This was the concept that struck Haldane, and from that came the idea that everything started when the first molecular *replicators* appeared on the primitive Earth.

Today, replication is firmly based on nucleic acids, but the nucleotides that make up these molecules are much more complex than the amino acids which produce proteins, as can be clearly seen in Figure 5.5. We know, in addition, that the abiotic synthesis of nucleic acids is far more difficult than that of proteins, and it is clear therefore that the structures required by replication were much more difficult to obtain in primitive conditions. This difficulty, however, could be overcome if it were shown that the first nucleic acids could have been preceded by simpler replicators, and a number of theories have been proposed precisely to that end. In 1982, for

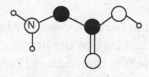

GLYCINE

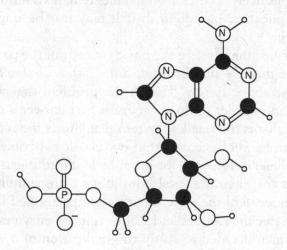

ADENOSINE MONOPHOSPHATE (AMP)

○ HYDROGEN   ◯ OXYGEN   Ⓝ NITROGEN   ● CARBON   Ⓟ PHOSPHORUS

Figure 5.5   Amino acids, the building blocks of proteins, are far simpler molecules than nucleotides, the building blocks of nucleic acids.

example, Cairns-Smith suggested that the first replicators were made of clay, i.e. that the first life forms were based on silicon. According to this theory, carbon-based replicators (the first RNAs) came later, and managed to propagate themselves by parasitising the silicon replicators, because they could never have done it alone. The processes suggested by Cairns-Smith do not exist in nature, today, but *could* have existed in the past, and may be given a small but finite probability.

A different solution has been proposed by Joyce, Schwartz, Orgel and Miller (1987) with the idea that the first nucleotides were simpler than modern ones, while Wächtershäuser (1992) has suggested that they were "tribonucleic acids", double-helical molecules that could have been formed on pyrite surfaces. Today there still is no satisfactory solution for the origin of nucleic acids, and the fact that they are objectively "difficult" molecules remains a serious obstacle for the replication paradigm, but it may not be impossible to overcome it.

Let us come therefore to the basic concept of the paradigm, i.e. to the idea that *the smallest replicative system is simpler that the smallest metabolic system*. This is the problem that we need to address, and in order to do so we must first answer a preliminary question: what is the smallest system that allows the replication of RNAs? The answer has come from two classic experiments, one by Sol Spiegelman in 1967 and the other by Manfred Eigen in 1971. In both cases the environmental conditions were simplified to the highest degree, and the experiments were performed in solutions containing free nucleotides and RNA-replicase enzymes.

Spiegelman decided to study the replication of a well-known nucleic acid, and chose the RNA of the virus Q_β, a chain of 4500 nucleotides that contains the coding instructions for all viral proteins, including the enzyme that replicates its own RNA. Spiegelman removed all proteins, put the viral RNA in solution, and observed that for a certain period it was faithfully replicated. Soon, however, a mutant appeared that by chance had lost the genes of some proteins that were not contributing to *in vitro* replication, and were therefore totally useless. Being shorter, the mutant was replicated faster and soon became more numerous than the original RNA. Then a second, shorter, mutant appeared which took the place of the first, and so on with other mutants, until the virtually complete elimination of all RNA pieces that were not taking part in replication. In the end, more than 90% of the initial RNA was lost, and all that remained in the test tube was a mutilated relic of 220 nucleotides (the *Spiegelman monster*) that could be replicated indefinitely.

Eigen repeated Spiegelman's experiment without introducing any RNA at the beginning, and limited himself to watching the evolution of a solution made of nucleotides and RNA replicase. This enzyme can form bonds between nucleotides, which made short nucleotide chains appear in the initial solution, and gradually these chains grew longer by the addition of other monomers. In the end, the solution reached a stationary state with RNA molecules that contained an average of 120 nucleotides.

The experiments of Spiegelman and Eigen arrived therefore at a very similar final result from two opposite starting-points. Either by cutting down a long RNA molecule, or by building up an increasingly longer molecule, the result was that the smallest nucleic acid system that was capable of indefinite replication was a solution of RNAs containing 100-200 monomers.

Dyson, it will be remembered, proved that the smallest metabolic system capable of maintaining itsef must contain at least 10 000 monomers of at least ten different types, and clearly such a system is far more complex than the smallest replicative system. We conclude therefore that the starting idea of the replication paradigm is fundamentally correct. Spiegelman and Eigen used a highly specific enzyme in their experiments, and this is not a realistic simulation of primitive conditions, but the theoretical conclusion that replication is simpler than metabolism is still valid. It remains to be seen, however, if the replication paradigm can really account for the processes that led to the origin of the first cells.

The RNA world

Between 1981 and 1986, Thomas Cech and Sidney Altman discovered that there are at least two metabolic reactions where the catalysts are not proteins but RNAs. Up until then, it had been accepted that all enzymes are proteins, and normally the discovery of a few exceptions does not undermine a virtually universal rule, but those two examples had an extraordinary implication.

In 1986, Walter Gilbert formulated it explicitly in these terms: *"If there are two enzymic activities associated with RNA, there may be more. And if there are activities among these RNA enzymes, or ribozymes, that can catalyse the synthesis of a new RNA molecule from precursors and an RNA template, then there is no need for protein enzymes at the beginning of evolution. One can contemplate an RNA world, containing only RNA molecules that serve to catalyse the synthesis of themselves."*

In reality, the theoretical possibility of an RNA world had already been suggested by Francis Crick with two prophetic statements. In 1966, Crick wrote that *"Transfer RNA looks like Nature's attempt to make RNA do the job of a protein"*, and in 1968 he added *"Possibly the first 'enzyme' was an RNA molecule with replicase properties."* Another precursor of the RNA world was H.B. White (1976), who noticed that *"many coenzymes are nucleotides, or use bases derived from nucleotides, and it is possible therefore that coenzymes are molecular fossils of the earliest RNA-based enzymes."* The discoveries of Cech and Altman suddenly gave a concrete basis to these ideas, and had an enormous impact because they were falling on fertile ground, already prepared to accept them. Two examples, however, are not enough to prove an hypothesis that concerns all molecules of a bygone primordial past. More precisely, the RNA world hypothesis requires a confirmation of three great generalizations:

(1) There has been a period in evolution when all genes were *ribogenes*.
(2) There has been a period in evolution when all enzymes were *ribozymes*.
(3) Modern RNAs are the remains of that primitive RNA world, and are therefore the most ancient molecules of the history of life, some

of which are still functional while others have become molecular fossils.

The experimental data that we have today are not yet conclusive, if taken one by one, but together they form a very strong case in favour of an early historical role of RNAs. There are five main facts that support this conclusion:

(1) RNAs can still function as genes (many viruses have genomes entirely made of RNA).

(2) RNA nucleotides are produced by direct syntheses, while DNA nucleotides are synthesised indirectly from ribonucleotides. This suggests that the first nucleic acids were RNAs, and that the mechanisms of their synthesis were later extended to DNA synthesis.

(3) In abiotic conditions it has been possible to obtain all four RNA bases (adenine, cytosine, guanine and uracil), but not the DNA characteristic base (thymine). *In vivo*, furthermore, thymine is synthesised from uracil-based precursors.

(4) The universal molecule that all cells employ as energy currency (ATP) is synthesised from RNA precursors.

(5) Many coenzymes (NAD, FAD, coenzyme A, etc.) contain ribonucleotides or bases derived from ribonucleotides (Benner *et al.*, 1989).

Collectively, these facts strongly suggest that RNAs had a leading role in what here has been called *postchemical* evolution. It must be underlined, however, that RNAs are *"sophisticated, evolutionarily advanced molecules"* (Miller, 1987; Joyce, 1989; Orgel, 1992), and all the above facts do not allow us to conclude that they were also leading players in the earlier phase of *chemical* evolution.

The fact that ribozymes came before protein enzymes does not mean that replication came before metabolism, but it is an historical fact that this is precisely the meaning that was given, almost universally, to the discovery of ribozymes. With very few exceptions, the RNA world has been interpreted with the logic of the replication paradigm: if RNAs could behave as genes and as enzymes, then they did it immediately, at the very beginning, and became the first replicators in the history of life.

According to this scheme, practically all cellular structures and functions came later, and only RNA replication was present at the beginning. This is clearly a *replication-first* hypothesis, and in fact it is the more updated version of the replication paradigm, but it does not follow directly from the experimental data. It is important therefore to check if it is really possible to say that the RNA world was a *world of replicators*.

Replication catastrophes

The idea that life started with RNA replicators implies that the primitive Earth became populated by RNAs that could either replicate themselves or could catalyse the replication of other RNAs. Up until now it has not yet been proved that RNAs can self-replicate, but it has been shown that some of them do have a weak replicase activity, and can therefore replicate other RNAs, thus forming a system that is, collectively, capable of replication. This allows us to conclude that some primitive RNAs could behave, to a certain extent, as the replicases used by Spiegelman and Eigen in their experiments, and that similar experiments were performed by nature, on a far larger scale, some 4 billion years ago.

The environmental conditions of the primitive Earth were surely different from those of Spiegelman's and Eigen's test-tubes, but this can be regarded as a secondary complication, and in a first approximation it can be ignored. What we cannot ignore, however, is the fact that any replication process is inevitably affected by errors, and it is therefore imperative to understand the consequences that such errors have for the very survival of a replicating system. This is a crucial problem for all replication-first theories, because it has been proved that *any self-replicating system can tolerate replication errors only below a critical threshold.* Above such a threshold, the system is overwhelmed by a runaway *error catastrophe*, and is inexorably condemned to collapse. This is a fundamental problem, and in order to address it we need first to quantify the critical threshold.

If a replicating system is described by N bits of information, and every bit is copied with an error probability ε, the total number of errors made in every generation is $N\varepsilon$. The systems that make many errors are more heavily penalized by the environment than those that make few errors, and this can be quantified by associating to any system a selection factor S (systems that make no error have a selective advantage S over those that make one, these have a selective advantage S over those that make two, and so on). The selective advantage is therefore a probabilty of surviving, and since information is proportional to the logarithm of probability, we can say that the selective advantage S given by the environment is equivalent to a transfer of information equal to $\log S$. With this premise we arrive at the following conclusion: a replicating system can avoid the error catastrophe, and survive, if the information that is losing at each generation ($N\varepsilon$) is less than the information ($\log S$) that is acquiring by the selective advantage, i.e. if

$$N\varepsilon < \log S \qquad (5.1)$$

Condition 5.1 is a very severe constraint because it means that a replicating system can increase N (i.e. can become more complex) only if its replication becomes more efficient. In practice, condition 5.1 requires that the average replication error ε be inversely proportional to N, and a system can therefore increase its complexity by an order of magnitude only if the replication errors decrease by an order of magnitude.

This conclusion allows us to explain the experimental results obtained by Spiegelman and Eigen: the RNA molecules of an *in vitro* replicating system can reach a length of 100-200 monomers, and no more, because longer molecules are overrun by the error catastrophe. The threat of this catastrophe of course also existed for primitive RNAs, whatever the environmental conditions, and we must therefore find out whether it was possible to avoid it. Eigen himself (1977) raised the problem and proposed a possible solution.

Let us assume that RNA molecules could grow in primitive solutions up to a maximum length of N nucleotides. These replicating systems could not, *individually*, grow beyond N, but Eigen was able to prove that things would be different for systems that contained different types of RNA. If the cycles of X individual systems could be combined into a single complex cycle – which Eigen called a *hypercycle* – one would have a supersystem of *XN* monomers that would maintain the same replication potential as the individual systems. Eigen proposed that the formation of hypercycles was the mechanism by which primitive replicating systems could increase the number of their components, and therefore their own complexity.

In 1981, however, Ursula Niesert proved that things are not so simple, because hypercycles too are under threat, and risk being swept away by at least three new types of catastrophes. A hypercycle can be destroyed by a selfish RNA that replicates faster without giving any contribution to the system (the *selfish gene catastrophe*). It can furthermore be destroyed by a mutant that skips one or more steps, and makes the hypercycle progressively shorter until it degenerates into a simple cycle (the *short-circuit catastrophe*). And finally it has been demontrated that a hypercycle with more than four cycles is intrinsically unstable and spontaneously tends to degenerate (the *population collapse catastrophe*). As we can see, hypercycles do not guarantee that primitive RNAs had a realistic probability of going on replicating themselves for millions of years.

Primitive RNAs, in conclusion, could certainly behave both as genes and enzymes, but this does not save the replication paradigm, because it cannot avoid the various catastrophes that necessarily affect all replicators.

Eigen's paradox

In 1981, Manfred Eigen proposed that the first replicators would have had a greater chance of surviving if they had been housed in small compartments instead of being freely diffusing in a solution. This is because compartments have an individuality, and individuals are the units on which natural selection can operate.

In practice, the compartment hypothesis is equivalent to saying that primitive RNAs were replicating themselves in the interior of microscopic vesicles, and that the membranes of those vesicles were also being replicated. All this requires a system of RNAs and lipids where some coordination exists between the two molecular families, and such a system could have appeared only after a process of chemical evolution. Its further development belongs therefore to the second phase of precellular evolution, the phase where the first mixed systems evolved into the first cells.

We conclude that the replication paradigm has not been able, so far, to account for chemical evolution, but could be valid for *postchemical* evolution, and this, it will be remembered, is also Dyson's hypothesis (*metabolism first, replication second*). Let us examine therefore the evolutionary potential of primitive vesicles containing RNAs that could behave both as genes and enzymes.

The replication paradigm requires that protein enzymes were not present at the beginning, and RNA replication was therefore performed by ribozymes. Some RNAs can in fact behave as polymerases and replicases, but they are far less efficient than the corresponding protein enzymes, and the accuracy of their replications was necessarily very low. The experimental measures, obtained from interacting coupled nucleotides, have shown that without protein enzymes the replication error ε cannot be less than 0.01, which means, from formula 5.1, that primitive RNAs could not have, as an order of magnitude, more than 100 nucleotides (Maynard Smith and Szathmáry, 1995).

A more accurate replication required protein enzymes, but the synthesis of these enzymes required a primitive translation apparatus, and therefore the presence of genes for such an apparatus. The smallest genome that is capable of coding for a replicase and a rudimentary translation apparatus is not known, but it has been estimated that a minimum of ten genes is necessary in order to keep translation errors within tolerable limits. The appearance of a primitive genome presents, however, two contrasting problems.

If the genes are linked in a single chromosome, the system contains an RNA molecule that cannot have fewer that 1000 nucleotides, but this is an order of magnitude greater than the critical threshold ($N = 100$), and the chromosome would be destroyed by the error catastrophe, unless protein enzymes are already present to raise the critical threshold.

If the genes are physically separated, on the other hand, they would behave like the components of a hypercycle, and would have to face the three catastrophes that threaten all hypercycles, unless protein enzymes are already present to prevent these catastrophes. We are therefore caught in a vicious circle: *in order to have protein enzymes it is necessary to have a large genome, but in order to have a large genome it is necessary that protein enzymes be already present.* This paradox, known as *Eigen's paradox*, has been outlined by John Maynard Smith (1983) and by Eörs Szathmáry (1989), and has turned out to be a formidable obstacle for all replication-first theories.

We are bound to conclude that the replication paradigm does not offer a plausible model even for *postchemical* evolution. Of course we cannot exclude that future discoveries might modify such a conclusion, but it would be necessary to discover, among other things, that primitive ribozymes were making replication errors comparable to those of protein enzymes, and this is extremely unlikely.

For the time being, therefore, the only reasonable conclusion is that a true replication mechanism appeared only at the end of precellular evolution, when the first cells came into being. Both *chemical* evolution and *postchemical* evolution, in other words, had to take place with metabolic systems that were able to tolerate errors, because only in this way could they be immune to the replication catastrophes.

Biological replication, despite its theoretical simplicity, was extremely difficult to achieve in practice, and became possible only with the evolution of a system that was sufficiently complex to withstand the error catastrophes. And the first system that did achieve that complexity level can rightly be regarded as the first living cell.

Abandoning the replication paradigm, in conclusion, does not mean underestimating the importance of replication: it only means that, to the best of our knowledge, biological replication could not have appeared at the beginning but only at the end of precellular evolution. And this means that the real problem of the origin of life is understanding the true origin of replication.

The ribotype theory

In 1981 I proposed the first model of *postchemical* evolution with *the ribotype theory on the origin of life*, and with the concept of *ribotype*, a term that indicates all RNAs and ribonucleoproteins of any organic system (Barbieri, 1981). Since ribonucleoproteins are advanced compounds, the name *ribosoids* was introduced to indicate all molecules made of RNA, or RNA and peptides, and the ribotype was also defined as the collective of all ribosoids of an organic system.

The scenario of the ribotype theory starts with the first organic systems that were capable of producing RNAs, be they Fox's microspheres, Wächtershäuser's vesicles or Dyson's minicells, and is compatible therefore with almost all metabolism-first models of chemical evolution. More precisely, it is compatible with all scenarios where primitive metabolic systems could grow, divide by budding or fission, and diversify with a generalised drift mechanism of the type described by Kimura and by Dyson. After this first phase of chemical evolution, the rest of precellular evolution took place, according to the ribotype theory, in a world of ribosoids, that is to say in a *ribotype world*. The difference between an RNA world and a ribotype world, may seem small, at first sight, but in reality it is enormous, because the RNA world implies a replication-first paradigm while the ribotype world belongs to a metabolism-first framework.

The ribotype theory was proposed before the discovery of ribozymes, but was also based on the idea that some RNAs can behave as enzymes. More precisely, the idea was that some primitive RNAs were similar to fragments of ribosomal RNAs, and could catalyse a peptide bond between any two amino acids. They were, in other words, *polymerising ribosoids*. The idea that the active players of protein synthesis are ribosomal RNAs, and not proteins, was proposed in 1970 by Carl Woese, and in those days it was only a speculation, but in 1991-1992 Harry Noller provided strong evidence that ribosomal RNAs can catalyse the formation of peptide bonds, and in 1998 Nitta and colleagues were able to prove it conclusively. Today we know therefore that the starting-point of the ribotype theory is valid: some primitive RNAs could have been polymerising ribosoids which started forming peptides and small proteins with random amino acid sequences.

This is the novelty that characterised *the first part of postchemical evolution*, and had at least one important consequence: the polymerising ribosoids allowed for the first time the production of peptides and small proteins *inside the system*, with endogenous syntheses, instead of importing these molecules from the outside. And this switch from *exopoiesis* to *endopoiesis* was an essential prerequisite for the development of a true *autopoiesis*.

In addition to polymerising ribosoids, precellular systems were producing many other types of ribosoids, and, for statistical reasons, most of these were devoid of any metabolic value. A few, however, could have more interesting properties and behave, for example, like ribozymes or transfer-like RNAs. The first part of *postchemical* evolution was therefore a simple continuation of the metabolic processes of chemical evolution, with the difference that precellular systems were now carrying RNAs in their interior, which means that both the players and the rules of metabolism were slowly changing.

Let us now come to *the second part of postchemical evolution*, the stage that was destined to lead to the origin of the first cells. It is in this stage that we must look for an answer to the problem that the replication paradigm has been unable to solve: *how did primitive systems manage to increase their complexity without being destroyed by error catastrophes?* The ribotype answer is based on three points. (1) Polymerising ribosoids could spontaneously form aggregates of high molecular weight by self-assembly. This was a formidable mechanism because it could easily produce compounds that had dimensions in the order of 1 000 000 Da, as proved by the fact that viruses and ribosomes still achieve these dimensions entirely with self-assemby processes.
(2) It is known that ribosomes of different species can contain very different proteins, and yet all function as ribosomes. In this case, *the same function does not require the same components*, but only structures that collectively belong to a wide family of molecules. The ribosome function could reappear in new generations of ribosoids even without replication of the same molecules, a mechanism

that I called *quasi-replication*. Primitive ribosoids could certainly exploit such a mechanism, because their modern descendants clearly demonstrate its empirical reality. Polymerising ribosoids could therefore become protoribosomes of high molecular weights even without mechanisms of exact replication.

(3) A system that contains heavy protoribosomes can avoid error catastrophes because *high-molecular-weight structures absorb thermal noise*, and are immune to a wide range of perturbations. This conclusion is based on a general engineering principle that Burks (1970) expressed in this way: *"There exists a direct correlation between the 'size' of an automaton – as measured roughly by number of components – and the accuracy of its function."* In the case of protein synthesis, this means that, in order to be precise, ribosomes must be immune to thermal noise and must therefore be heavy.

This then is the solution of the ribotype theory: in order to avoid the error catastrophes in the journey toward exact replication, it was necessary to have high molecular weight protoribosomes, and the production of these ribosomes for an indefinite number of generations was possible, before exact replication, because ribosoids could achieve it with processes of self-assembly and quasi-replication. The development of high-molecular-weight protoribosomes took place during *postchemical* evolution, simply because all necessary conditions existed in that period, and the development could be realised with processes that were both natural *and primitive*.

Together with polymerising ribosoids, ribogenes were also evolving, which means that precellular systems had genomes predominantly made of RNAs. The dimensions of ribogenomes could also increase, because the mechanism of quasi-replication was avoiding the error catastrophes, and eventually ribogenomes and protoribosomes became sufficiently large for both to be immune to thermal noise. At that point, quasi-replication could be turned into exact replication, and precellular evolution came to an end.

The first cells which appeared on Earth, in conclusion, had high-molecular-weight protoribosomes, a mixture of ribozymes and protein enzymes, and RNA genomes. They truly were *ribocells*, and their appearance marked, to all effects, the origin of life.

The genetic code

Genes can replicate and transmit their linear information to other genes, but proteins cannot. The information of an amino acid chain is always coming from the information of a nucleotide sequence carried by a messenger RNA, and in this process an amino acid is always specified by a group of three nucleotides that is called a *codon*.

The messenger RNAs can be regarded therefore as ideally divided into nucleotide triplets, and since the combinations of four nucleotides in groups of three are 64 (4^3), there can be a total of 64 codons for 20 amino acids. The rules of correspondence between the 20 natural amino acids and the 64 codons represent, collectively, the *genetic code* (Figure 5.6).

As we can see from the figure, three codons are used as protein synthesis termination signals, while the other 61 specify the amino acids and the initiation signal. Between 61 codons and 20 amino acids there cannot be a one-to-one correspondence, and in fact some amino acids are specified by six codons, some by four, others by two, and only two amino acids are coded by a single codon. In technical terms, this is expressed by saying that the genetic code is *degenerate*.

The biological meaning of this degeneracy is one of the few properties of the genetic code for which we do have a rational explanation. The degeneracy could have been avoided by choosing, for example, 20 codons for 20 amino acids, three termination codons, and 41 nonsense codons. In this case, however, the chance mutation of a nucleotide would have produced in most cases a nonsense codon, and this would have interrupted protein synthesis. The great majority of mutations would not be expressed, and this explains why it is imperative that all 64 codons have a meaning. The degeneracy of the genetic code, in short, is necessary to ensure the expression not only of genes but also of all their possible mutations.

It is much more difficult, however, to understand why nature chose the genetic code that we have, and not one of the many other possible versions. This remains a mystery, but it is instructive to speculate on what could have happened if other codes had been chosen.

Let us imagine, for example, a code that has 20 codons for the

THE GENETIC CODE

Amino acids	Codons

Initiation signal and

Methionine	AUG
Tryptophan	UGG
Aspartic acid	GAC,GAU
Glutamic acid	GAA,GAG
Asparagine	AAC,AAU
Cysteine	UGC,UGU
Phenylalanine	UUC,UUU
Glutamine	CAA,CAG
Histidine	CAC,CAU
Lysine	AAA,AAG
Tyrosine	UAC,UAU
Isoleucine	AUA,AUC,AUU
Alanine	GCA,GCC,GCG,GCU
Glycine	GGA,GGC,GGG,GGU
Proline	CCA,CCC,CCG,CCU
Threonine	ACA,ACC,ACG,ACU
Valine	GUA,GUC,GUG,GUU
Argine	AGA,AGG,CGA,CGC,CGG,CGU
Leucine	CUA,CUG,CUG,CUU,UUA,UUG
Serine	AGC,AGU,UCA,UCC,UCG,UCU

Termination signals   UAA,UAG,UGA

(A=Adenine,C=Cytosine,G=Guanine,U=Uracil)

Figure 5.6   The genetic code which actually exists in almost all living organisms.

amino acids and 44 termination codons (Figure 5.7). Such a code does not contain nonsense codons, and all protein syntheses would be completed. Because of the high number of termination codons, however, the average length of proteins would have been considerably shorter than the average that exists in nature. That code, in other words, would have generated a *world of miniproteins*.

The opposite effect would have been produced by a code which contained only one termination codon. In this case the average protein length would have been three times the natural length, and would have produced a *world of maxiproteins*.

Let us consider now a code where aspartic acid and glutamic acid are specified by eight codons instead of four (Figure 5.8). Amino acids are divided into four great groups (acidic, basic, hydrophilic and hydrophobic) by the chemical properties of their side chains, and the two amino acids in question are the sole representatives of the acidic group (Figure 5.10). A code which had doubled their frequency, therefore, would have generated a *world of more acidic proteins*. Histidine and lysine, on the other hand, are basic amino acids, and if their frequency had been doubled, the result would have been a *world of more basic proteins*.

And finally let us come to the most equilibrated of all codes, the one in which every amino acid is specified by three codons (Figure 5.9). In this case we cannot figure out what kind of world would have been generated, but surely it would have been different from ours, because proteins would have had different statistical mixtures of amino acids and therefore different chemical properties.

All this tells us that the evolution of primitive ribosoids into protoribomes and ribogenomes could have produced – *at equal thermodynamic conditions* – a countless number of other protein worlds, and therefore countless other forms of life. In the course of precellular evolution, therefore, two distinct processes went on in parallel: the development of metabolic structures, and the development of *a particular* genetic code that gave life the familiar forms of our world, and not those of countless other possible worlds.

CODE A

Amino acids	Codons

Initiation signal	ACG
Methionine	AUG
Tryptophan	UGG
Aspartic acid	GAC
Glutamic acid	GAA
Asparagine	AAC
Cysteine	UGC
Phenylalanine	UUU
Glutamine	CAA
Histidine	CAC
Lysine	AAA
Tyrosine	UAC
Isoleucine	AUA
Alanine	GCA
Glycine	GGA
Proline	CCA
Threonine	ACA
Valine	GUA
Argine	AGA
Leucine	CUA
Serine	AGC
Termination signals	UAA,UAG,UGA,UGU,UCU,UUC
	UCC,UAU,GUU,UCA,UUG,UUA
	UCG,AUU,AUC,AAG,AAU,AUC
	ACC,AGG,AGU,CUG,CCG,GAU
	GCG,GAG,GGG,GCC,GUG,GGC
	GCU,GGU,GUC,CAG,CUU,CAU
	CGU,CCC,CGC,CUC,CCU,CGA,CGG

Figure 5.7   An hypothetical genetic code in which every amino acid is codified by a single codon. The high number of termination codons would give rise to a world of miniproteins.

CODE B

Amino acids	Codons

Initiation signal and

Methionine	AUG
Tryptophan	UGG
Aspartic acid	GAC,GAUGCU,CGU
Glutamic acid	GAA,GAG,GGU,CGG
Asparagine	AAC,AAU
Cysteine	UGC,UGU
Phenylalanine	UUC,UUU
Glutamine	CAA,CAG
Histidine	CAC,CAU
Lysine	AAA,AAG
Tyrosine	UAC,UAU
Isoleucine	AUA,AUC,AUU
Alanine	GCA,GCC,GCG
Glycine	GGA,GGC,GGG
Proline	CCA,CCC,CCG,CCU
Threonine	ACA,ACC,ACG,ACU
Valine	GUA,GUC,GUG,GUU
Argine	AGA,AGG,CGA,CGC,CGG
Leucine	CUA,CUG,CUG,CUU,UUA,UUG
Serine	AGC,AGU,UCA,UCC,UCG,UCU

Termination signals UAA,UAG,UGA

Figure 5.8   An hypothetical genetic code in which aspartic acid and glutamic acid are coded for by twice the number of codons that exist in the natural code, thus producing a world of more acidic proteins.

CODE C

Amino acids	Codons

Initiation signal	ACG
Methionine	AUG,AUU,ACU
Tryptophan	UGG,UUG,UUA
Aspartic acid	GAC,GAU,GCG
Glutamic acid	GAA,GAG,GGG
Asparagine	AAC,AAU,UCG
Cysteine	UGC,UGU,UCU
Phenylalanine	UUC,UUU,UCC
Glutamine	CAA,CAG,CUU
Histidine	CAC,CAU,CGU
Lysine	AAA,AAG,UCA
Tyrosine	UAC,UAU,GUU
Isoleucine	AUA,AUC,GGU
Alanine	GCA,GCC,GCG
Glycine	GGA,GGC,GCU
Proline	CCA,CCC,CCG
Threonine	ACA,ACC,CGA
Valine	GUA,GUC,CGG
Argine	AGA,AGG,CUG
Leucine	CUA,CUC,CCU
Serine	AGC,AGU,CCG
Termination signals	UAA,UAG,UGA

Figure 5.9   An hypothetical genetic code in which all amino acids are codified by three codons. It would be the most balanced code, in the sense that all amino acids would have the same statistical frequency, but it has not been nature's choice.

Families	Amino acids	Codons
1 ACIDIC	Aspartic acid	GAC,GAU
	Glutamic acid	GAA,GAG
2 BASIC	Histidine	CAC,CAU
	Lysine	AAA,AAG
	Arginine	AGA,AGG,CGA, CGC,CGG,CGU
3 HYDROPHILIC	Asparagine	AAC,AAU
	Glutamine	CAA,CAG
	Tyrosine	UAC,UAU
	Threonine	ACA,ACC,ACG,ACU
	Serine	AGA,AGU,UCA, UCC,UCG,UCU
4 HYDROPHOBIC	Methionine	AUG
	Tryptophan	UGG
	Phenylalanine	UUC,UUU
	Cysteine	UGC,UGU
	Isoleucine	AUA,AUC
	Alanine	GCA,GCC,GCG,GCU
	Glycine	GGA,GGC,GGG,GGU
	Proline	CCA,CCC,CCG,CCU
	Valine	GUA,GUC,GUG,GUU
	Leucine	CUA,CUC,CUG,CUU, UUA,UUG

Figure 5.10   The natural amino acids are divided into four families according to their electric charge and to their reactivity with water.

Evolution of the code

The genetic code is one of the most universal structures of life, and even its rare variant forms, which exist in some micro-organisms, are different only in a few minor details. Such extraordinary uniformity means that the rules of the code have not been changed during the history of life, and go all the way back to the very origin of the cell. Which is understandable, because a change in the code would have changed the structure of all proteins, and the entire system would have collapsed. A fully functioning code, however, emerges only when all its rules are present, and we need to understand what existed before that point. Can we conceive half a code or a quarter of a code? And what sense would half a code have made before the origin of life, when there was no exact replication to pass its rules on?

The simplest way of answering these questions is by discussing a thought experiment, a highly idealised example that allows us to focus on the essential points of the problem. Let us imagine a primitive system which had simplified versions of the three main protagonists of protein synthesis, i.e. preribosomes, transfer-like RNAs and messenger-like RNAs. The messenger-like molecules were not transporting any message, and were mainly random linear sequences of nucleotides, but could still have a role to play. We have seen that most metabolic reactions are more likely to occur on a surface than in space, and I suggest that *the same thermodynamic arguments tell us that they are even more likely to occur along a line than on a surface.* A linear chain of RNA was allowing frequent encounters of preribosomes and transfer-like RNAs, and amino acids could be brought close together more easily on a line than on a surface. Thermodynamics, in other words, was favouring systems in which peptide bonds were made by preribosomes and tRNAs attached to linear chains of RNA.

We can easily imagine a period when no code was existing, and any tRNA could bind to any amino acid irrespective of its anticodon. In these conditions, peptide bonds were made at random, and only *statistical* proteins were synthesised. Let us suppose now that one of the tRNAs underwent a change that allowed it to bind only one particular amino acid. This is equivalent to the appearance of a single rule of the genetic code, but the effect would still have been dramatic. If the amino acid in question had been hydrophilic, for example, all proteins of the system would have become more hydrophilic; if it had been hydrophobic, acidic or basic, the whole system would have become more hydrophobic, more acidic or more basic, and so on.

The statistical proteins of these systems did not yet function as enzymes (this role was still performed by ribozymes) but could provide structural support, could modify the local microenvironment, and above all were maintaining the precellular systems in a state of continuous metabolic activity that was essential for the development of a true autopoiesis. The appearance of a few coding rules would not have created, in these conditions, any specific property, but surely would have produced new types of stastistical proteins and new physico-chemical conditions, with the result that the system would have become more heterogeneous. Different linear RNAs would have started favouring the synthesis of different groups of statistical proteins, thus becoming increasingly similar to messenger RNAs. In this way, the individuality of messenger RNAs would have emerged gradually, like a photographic image which is gradually brought into focus.

We find a similar process in the evolution of linguistic codes. The sounds uttered by the first speakers were probably little more than random combinations of vowels and consonants, at the beginning; then they were divided into a few major categories (sounds of friendship, enmity, fear, satisfaction, etc.), and finally they managed to express an increasing number of meanings. The evolution of the rules went on hand in hand with the evolution of the words, and the two processes, although intrinsically different, evolved in parallel.

We conclude that, during postchemical evolution, what was taking place was not only a development of metabolic structures, but also an evolution of coding rules, of *natural conventions*. The true mechanism of postchemical evolution, in other words, was not genetic drift alone, but a combination of drift and natural conventions. To the classical concepts of evolution by genetic drift and by natural selection, we must add therefore the concept of *evolution by natural conventions*.

It is important to notice that this is very different from the mechanism of chemical evolution. Kauffman and Dyson, it will be remembered, have shown that the probability of a spontaneous transition from chaos to order increases with the complexity of the system, but in this case the order (or antichaos) is not a result of natural conventions and has nothing to do with organic codes.

This tells us that *chemical evolution was really different from postchemical evolution*. In the course of chemical evolution, the jump of primitive metabolic systems from chaos to order was only a question of statistical probability and energy conditions. During postchemical evolution, instead, a new type of antichaos appeared, an order that was based on conventional rules of correspondence between two independent molecular worlds, and it was from these first natural conventions that the genetic code finally emerged.

The ribotype metaphor

The RNAs and the ribonucleoproteins of a living system have been collectively defined as the *ribotype* of the system (Barbieri, 1981). In the paper which introduced that concept, however, it was explicitly stated that the new term has also a deeper meaning, because it represents *a new cell category*. If the word *ribotype* were used only to indicate a class of molecules, we could call *glycotype* the carbohydrates, or *lipotype* the fatty acids, but we would not have new categories. Sugars, fats and proteins, in fact, all take part in cell metabolism and belong to the same category, i.e. to the phenotype. The ribotype, on the contrary, has a biological role that is qualitatively different from those of the two traditional categories. As phenotype is the seat of metabolism and genotype the seat of heredity, so ribotype is the seat of genetic coding. *The distinction between phenotype, genotype and ribotype reflects the distinction between energy, information and meaning*, the most fundamental of nature's entities.

The 1981 paper, furthermore, pointed out that the ribotype is not only independent from genotype and phenotype, but has a logical and a historical priority over them. According to the ribotype theory, primitive ribotypes developed a *ribophenotype* (the ribozymes) and a *ribogenotype* (the ribogenes), and these last two categories evolved into what we now call phenotype and genotype.

We are however accustomed to think of RNAs and ribonucleo-proteins as *intermediaries* between genotype and phenotype, and it is difficult to regard intermediaries as being as important as the quantities they connect. The logical difficulty that we face, with the ribotype concept, is even more general, because we have been taught to approach our problems in terms of *dichotomies* (nature–nurture, mind–body, heredity–environment, genotype–phenotype, etc.). And the origin of life is universally approached with the mother of all dichotomies: the *chicken-and-the-egg* metaphor. The real protagonists of life, we are told, are genes and proteins, and the problem of the origin is understanding *whether it was the chicken or the egg that came first.*

In this framework, a theory based on *three* categories is totally out of place, and so we have no choice but to go straight to the heart of the matter: can we replace the *chicken-and-the-egg* riddle with a better metaphor? As a matter of fact, as soon as we take a closer look at that time-honoured analogy we realise that there is something wrong with it. The egg and the chicken are not the two faces of one dual system. They are two dualistic systems in different stages of development. Each one of them is a complete genotype–phenotype entity, and it is pure fiction to say that one represents the genotype and the other stands for the phenotype.

We do indeed need a better metaphor, and luckily there is one at hand. It is the metaphor of the *cell-as-a-city*, where the proteins of a cell are compared to the houses of a city, and the genes to their blueprints (Barbieri, 1981, 1985). In this framework it would not even make sense to ask if it was the houses or the blueprints that came first. What came first was a third party, the inhabitants, i.e. the intermediaries between houses and blueprints in a city which correspond to the intermediaries between proteins and genes in a cell.

Despite its intuitive appeal, the *cell-as-a-city* metaphor has not become anything like as popular as the *chicken-and-the-egg*, and it is highly instructive to understand why. The crucial point is that in a city only the inhabitants are alive, whereas houses and blueprints are not. The city metaphor, in other words, implies that *genes and proteins are molecular artifacts, just as blueprints and houses are human artifacts*. This is the preposterous idea. How can one accept that genes and proteins, the very molecules of life, are inanimate manufactured objects? That probably explains why the ribotype theory has not attracted the attention of the origin-of-life people. And yet it has never been proved that the preposterous idea is false. It may therefore be worth taking a closer look at it.

Copymakers and codemakers

There was a time when atoms did not exist. They came into being within giant stars, and were scattered all over the place when those stars exploded. There was a time when molecules did not exist. They originated from the combination of atoms on a variety of different places such as comets and planets. There was a time when polymers did not exist. They were produced when molecules joined together at random and formed chains of subunits. There was a time when all the polymers of our planet were random polymers, but that period did not last forever. At a certain point, new types of polymers appeared. Some molecules started making copies of polymers, and for this reason we will call them *copymakers*. Other molecules made coded versions of the copies, and we will refer to them as *codemakers*. On the primitive Earth, the copymakers could have been RNA replicases and the codemakers could have been transfer RNAs, but other possibilities exist, and so here we will use the generic terms of copymakers and codemakers. The only thing that matters, for our purposes, is the historical fact that copymakers and codemakers came into being and started producing copied molecules and coded molecules.

Now let us take a look at these new polymers. The formation of a random chain of subunits is accounted for by the ordinary laws of thermodynamics and does not require any new physical quantity. But when a copymaker makes a copy of that chain, something new appears:

the sequence of subunits becomes *information* for the copymaker. In a similar way, when a codemaker takes a chain of monomers of one kind to produce a chain of monomers of a different kind, something new appears: the second chain becomes the *meaning* of the first one. *It is only the act of copying that creates information, and it is only the act of coding that creates meaning.* Information and meaning, in other words, appeared in the world when copymakers and codemakers came into existence and started functioning.

The appearance of copied polymers and coded polymers was a major event also for another reason. Up to that point, all molecules formed on the primitive Earth had one thing in common: their structure was entirely determined by the assembly properties of their atoms, i.e. *from within*. In the case of copied and coded polymers, in contrast, the order of the subunits was determined by external templates, i.e. *from without*. In everyday language, we distinguish between *natural* and *artificial* products in a straightforward way: the objects that are formed spontaneously are natural, while those which are shaped by external agents are artificial. And that is precisely the distinction that exists between random polymers on one hand and copied or coded polymers on the other. I conclude therefore that copied molecules (genes) and coded molecules (proteins) are indeed, in a very deep sense, *artificial* molecules. They are artificial because they are produced by external agents, because their primary structure is determined from without and not from within, because their production involves outside processes based on information and meaning.

There was a time when the world was inhabited only by *natural* molecules, but that period did not last forever. At a certain point copied and coded molecules appeared, and the world became also inhabited by *artificial* molecules, by *artifacts made by nature*. And that was not just another step toward life. It was the appearance of the very logic of life because, from copymakers and codemakers onward, all living creatures have been artifact-makers. In a very fundamental sense, we can define life itself as *artifact-making*.

The handicapped replicator

The cell-as-a-city metaphor suggests that proteins and genes are artificial molecules, and we have just seen that, deep down, that is precisely what they are. The metaphor also suggests that modern cells are to primitive cells what large cities are to small villages, and this is not an unreasonable analogy. Modern eukaryotic cells, for example, contain millions of ribosomes, like the inhabitants of large cities, while prokaryotic cells have only hundreds or thousands of ribosomes, like the inhabitants of villages.

The metaphor can also be extended to earlier stages of evolution. If the origin of the first cells is likened to the origin of the first villages, we can compare the age of precellular evolution to the period of history in which villages did not exist. The interesting point is that this metaphor allows us to take a closer look at today's most popular model on precellular evolution: the model of the naked gene as the first replicator (Dawkins, 1976).

Dawkins has readily admitted that genes are not doing any replication, but since they code for the molecules that replicate them, he finds it legitimate to call them *replicators* in order to avoid long periphrases. Michael Ghiselin (1997) has pointed out that this is confusing the "object" with the "agent" of replication, but Dawkins's use of the word has stuck, and today most biologists seem to take for granted that genes are replicators. This is why I have avoided that word altogether and I have used the term *copymakers*. The distinction between copymakers and copies is still alive and well, and so there is no danger of confusing what is copied with what does the copying. Whatever one's choice of words, however, the real point is the substance, not the terminology.

The substance of the replicator model is that *all that matters in life is information, and all that matters in evolution is the replication of information with occasional mistakes*. But we have seen that at the heart of life there are two fundamental entities, not one. Information and meaning are two independent entities, copying and coding are two independent processes, and the codemaker between genes and proteins must be a third party because otherwise there would be no real code. The replicator model is not wrong, but incomplete (or handicapped), because what matters in life is *replication and coding*, not replication alone (I prefer to speak of *copying and coding*, but the message is the same). The replicator model would be right if the cell were a Von Neumann automaton where the hardware is completely described by the software, and information is really everything, but nature has not taken that path. And probably for very good reasons, because that path was seriously undermined by the error catastrophes.

One could still argue, however, that a "naked gene" phase should have preceded a phase of "copying-and-coding", and this is where the cell-as-a-city metaphor can help us. The metaphor suggests that before cities there were villages, that before villages there were humans living in the open, that before humans there were ancestral hominids, and so on. The point is that in all stages there were "agents" not just "objects". There has never been a time in precellular evolution in which copied molecules (genes) could exist without copymakers, or coded molecules (proteins) without codemakers. It was copymakers and codemakers that came first, because it was they who were the first "agents" in the history of life. The first molecules of the ribotype world were produced by random processes and the chances of getting copymakers or codemakers (for example, RNA replicases or transfer RNAs) were not substantially different. Any one of them could have appeared before the other, without making much difference. What did make a difference was the appearance of both ribosoids because only their combination created a renewable link between genes and proteins. It was a ribotypic system containing copymakers and codemakers that started life, because that was the simplest possible *lifemaker*, i.e. the simplest possible *agent*. Admittedly, a naked gene would have been a simpler system, but it would not have been an *agent*, and that makes all the difference. As Einstein once remarked, *"Things should be made as simple as possible, but not simpler."*

6

PROKARYOTES AND EUKARYOTES

After the origin of life, our planet has been inhabited only by single cells for some 3000 million years, i.e. for more than 85% of the entire history of life on Earth. Multicellular creatures appeared only at the end of that unimaginably long period, but when they did they quickly diversified into all known kingdoms. This chapter presents a brief account of what we can reconstruct about *the age of the cell*, and its conclusions are not always in line with present thinking. It is shown, for example, that the *bacteria-first* view is unlikely to be correct, despite it persistent popularity. Another minor unorthodox conclusion is that the kingdoms of life are probably seven, and not five or six as reported in most textbooks. The major divergence from orthodoxy, however, comes from the fact that cellular evolution is reconstructed by taking organic codes into account. The emergence of the cell nucleus, for example, is related to the full development of the splicing codes, because it is these codes that allow a physical separation, in space and time, between transcription and translation. There are also good arguments for the existence of *cytoskeleton codes* and *compartment codes*, thus suggesting that these were instrumental to the evolution of other major eukaryotic structures. The idea that organic codes have something to do with the great events of macroevolution, in brief, does not seem unreasonable, and gives substance to the concept of *evolution by natural conventions*.

The potassium world

Life was born in the sea, and even the organisms that invaded the land could do so only by carrying with them an *internal sea* that enabled their cells to continue to live in water. This liquid that floods every cell still has values of pH and osmotic pressure which are similar to those of sea water, and likewise contains high concentrations of sodium and potassium ions. The really extraordinary thing, however, is that inside all cells (including those that live in the sea) the concentrations of sodium and potassium are totally different from those of the surrounding liquid.

A first explanation of this strange experimental fact came into view when it was found that sodium ions (Na^+) produce very high osmotic pressures inside the cell, and, without defence mechanisms, a cell would swell to bursting point and nothing could save it. The osmotic defence mechanisms can be of three kinds: (1) an external wall that prevents swelling from the outside, (2) an internal web of filaments that ties the cell membrane from the inside, or (3) a battery of ion pumps on the cell membrane which contually drains out the excess sodium.

For a long time it has been thought that the cell's osmotic problems are caused by sodium alone, but things turned out to be more complicated. The real problem is that cells need very high concentrations of potassium ions (K^+) in their interior, and to this end they are continually importing potassium from the outside. The cell can cope with potassium osmotic pressure, but not with much higher values, and it is for this reason that the addition of sodium ions would be lethal. Hence the need for sodium defence mechanisms. As we can see, the need to counteract the osmotic effects of sodium comes primarily from the cell's vital need to maintain very high concentrations of potassium in its interior, and this brings us to the real biological problem: why do cells need potassium so much?

The first clue was provided by Martin Lubin (1964) and by Cahn and Lubin (1978), with the discovery, first in bacteria and then in eukaryotes, that protein synthesis comes to a halt if high concentrations of potassium ions are not present. As a matter of fact, potassium is also required by many other metabolic reactions, but protein synthesis is so important that that alone can account for the fact that cells are totally dependent on potassium. It has been shown, however, that potassium ions do not usually take part in metabolic reactions, and mainly contribute to setting up the chemical environment for them. Which brings us back to the key question: why do cells depend so much on potassium?

The simplest explanation is that the protein synthesis machinery started evolving in the presence of potassium, and, once set up, the mechanism could no longer be changed. This is the same argument which has been used to explain the conservation of the genetic code. Potassium dependence and the genetic code involve so many interdependent things that any change would make the whole system collapse.

At this point, however, we cannot ignore the fact that the evolution of protein synthesis started before the origin of the first cells, in systems which could not have cell walls, cytoskeleton filaments or sodium pumps, for the very good reason that all these structures require well-developed proteins. How could precellular systems have high potassium concentrations, and low sodium levels, without any of the molecular mechanisms that cells employ to this end? The most plausible answer is that those concentrations did not have to be produced in prebiotic systems because *they already existed in the environment of the primitive seas.* The ribotype world, in short, was also *a potassium world.*

It is not possible, of course, to rule out other explanations, but none has the simplicity and the explicative power of this direct environmental hypothesis. And we cannot exclude the possibility of an experimental test: it would be enough to discover in the geological record the signs of ancient seas where potassium was more abundant than sodium (such a discovery would allow us, among other things, to put a date on the origin of life).

The potassium world hypothesis does not seem to have a clear paternity. It was circulating in the 1970s at informal meetings on the ionic conditions of protein synthesis, but I have been unable to trace it to a precise source. Today it is less popular, but it is mentioned here because it could have important consequences for the first stages of cellular evolution. If the hypothesis is right, we have to conclude that the ancient potassium seas gradually turned into sodium-dominated oceans, and this slow but deep transformation of the planet (comparable to the global change of the Earth's atmosphere by the introduction of oxygen) must have had an influence on the evolution of the first cells.

Two forms of life

The greatest divide of the living world is not between plants and animals, as was thought for thousands of years, but between cells without a nucleus (*prokaryotes*) and nucleated cells (*eukaryotes*). Prokaryotes, or *bacteria*, have only one DNA molecule, arranged in a circle, and a single cytoplasmic compartment where all biochemical reactions take place in solution, and normally the form of the cell is due to an external wall (an exoskeleton) which surrounds the cell's plasma membrane.

Eukaryotes have various DNA molecules, arranged in linear fibers which are repeatedly coiled and folded to produce highly organised chromosomes, and a composite cytoplasm which is divided into distinct compartments and houses a variety of cell organelles (mitochondria, chloroplasts, lysosomes, the endoplasmic reticulum, etc.); the form of the cell is due to an internal cytoskeleton which is made of three different types of filaments (microtubules, microfilaments and intermediate filaments).

These structural differences are but a reflection of two very different lifestyles. Prokaryotes live almost exclusively as single cells, and can inhabit virtually any ecological niche, with or without light, with or without oxygen, with or without organic molecules. Most of them are capable of synthesising all their components from inorganic molecules and an energy source, can rapidly adapt to environmental changes, and exchange genes in a horizontal way, i.e. between individuals of the same generation. Bacteria, furthermore, are apparently capable of avoiding extinction, and some living forms appear strikingly similar to those found in fossils which are more than 3 billion years old.

Eukaryotes, by contrast, are mostly oxygen-dependent organisms (many are obligate aerobes), and, in addition to monocellular forms (*Protozoa* or *Protista*), have generated all three kingdoms of multicellular beings (plants, fungi and animals). They invented new mechanisms of cell division (mitosis and meiosis), new types of movement, meiotic sexuality and above all embryonic development, a process that is potentially capable of generating countless different structures. In general, eukaryotes can adapt to the environment with highly sophisticated anatomical changes, but the price of their versatility is a very high level of extinction (the average lifetime of animal species is only 4 million years).

Carlile (1980) has analysed the difference between bacteria and eukaryotes in terms of what ecologists call *r*-strategies and *K*-strategies. Bacteria are supreme *r*-strategists, in the sense that they multiply rapidly when resources are abundant, and react to environmental changes with fixed mechanisms that appear to have remained invariant for millions of years. Eukaryotes, by contrast, are mainly exploiting *K*-strategies of survival which allow them to make efficient use of scarce resources, and in general can overcome severe crises only by inventing new technological solutions. Prokaryotes and eukaryotes, in short, are not only different types of cells, but two radically different forms of life, and we need to understand how such a deep dichotomy could have originated.

In 1866, Ernst Haeckel proposed a phylogenetic tree where the first forms of life were cells without a nucleus (which he called *Monera*), which later generated nucleated cells (*Protista*), which in turn gave rise to all multicellular organisms. Already in 1883, Schimper proposed that chloroplasts had once been free-living bacteria that happened to be incorporated, by a kind of symbiosis, into some eukaryotes, and from 1905 to 1930 this hypothesis was not only reproposed but also extended to mitochondria by Mereschowsky, by Portier and by Wallin.

In the 1970s, the symbiosis hypothesis was forcefully reproposed by Lynn Margulis, and within a few years it received the support of an astonishing number of experimental discoveries. Mitochondria and chloroplasts are still carrying fragments of their ancient circular DNA, and have 70S ribosomes which are typical of bacteria, all of which leaves little doubt about their origin. It is practically certain, therefore, that mitochondria and chloroplasts were acquired by symbiosis during cellular evolution, and paleontology even allows us to establish that this happened around 1500 million years ago, because it is only after that period that geological strata show fossilised cells that are large enough to contain intracellular organelles.

All this, however, tells us nothing about the cells that acquired orgenelles by symbiosis, and on this point biologists are divided into two opposing camps. Some maintain that the cells which engulfed bacteria were themselves bacteria, a hypothesis which leads to two precise conclusions: (1) the first living cells were bacteria, and (2) eukaryotes are chimeras of bacteria (*the bacterial theory of life*).

Other biologists are convinced that the acquisition of bacteria by symbiosis required characteristics that do not exist in bacteria, which means that both the direct ancestor of eukaryotes and the common ancestor of bacteria and eukaryotes could not have been bacteria. The disagreement, in conclusion, is about two major points: (1) the nature of the first cells, and (2) the origin of eukaryotes. It is important therefore to understand whether the first cells which appeared on Earth were bacteria, as maintained by the bacterial theory of life, or cells which had the potential to generate both bacteria and ancestral eukaryotes.

Three primary kingdoms

All living cells contain ribosomal RNAs, and it seems that these nucleic acids have changed very little in the course of evolution because their structures are similar in all organisms. Despite this enormous molecular uniformity, however, all species are slightly different in their ribosomal RNAs, and in 1977 Carl Woese showed that these little differences give us precious information on the very first stages of cellular evolution.

Woese chose to study the RNAs of the small ribosomal subunits: having extracted and purified these long molecules, he cut them at various points with enzymes, and then classified the fragments according to the number and the types of their nucleotides. The procedure was applied to the RNAs of different species, and in this way Woese obtained tables of nucleotide fragments (called *matrices of association coefficients*) where the differences among species could be expressed in a quantitative way.

It was like comparing different languages by cutting their words into groups of letters, and by classifying these groups according to the number and the types of their letters. Such an experiment has actually been performed, and has made it possible not only to calculate the "distance" that exists between any two languages, but also to divide languages into families on the basis of such distances. A family contains all languages which have small distances among them and big distances with the others. For example, Latin languages (Italian, French, Spanish and Romanian) form a family which is quantitatively different from the Germanic family (English, German, Dutch and Swedish) as well as from the Slavic family (Russian, Polish, Czech and Bulgarian). Woese demonstrated that ribosomal RNAs allow us to calculate in a similar way the distances that exist between species, and found that this divides all cells in three major groups: *eubacteria, archaebacteria* and *eukaryotes.*

This discovery has two outstandingly important implications:
(1) Bacteria do not form a monophyletic group but two distinct kingdoms (eubacteria and archaebacteria).
(2) The distance between the two bacterial kingdoms is comparable to the distance that divides each of them from the eukaryotes' ancestors (cells that Woese called *urkaryotes*).

This means that the three groups of cells have the same phylogenetic antiquity and represent three distinct taxa at the highest level (*three primary kingdoms*), none of which can be ancestral to the others. Unfortunately, the phylogenetic distances obtained from molecular data do not represent real times, and therefore do not tell us when the separation between the three primary types of cells did take place. They tell us, however, that the "birthdate" of the eukaryotic kingdom was as old as that of the bacterial kingdoms, and this in turn suggests that there was a common ancestor of all three primary kingdoms, an ancestral cell that Woese called *progenote*.

In reality, the distances obtained by Woese are compatible with four different phylogenetic trees (Figure 6.1), i.e. with models where the three kingdoms are equally old (Figure 6.1A), where archaebacteria are nearer to eubacteria (Figure 6.1B), where eubacteria are nearer to eukaryotes (Figure 6.1C), or, finally, where archaebacteria are nearer to eukaryotes (Figure 6.1D). In the first case (Figure 6.1), the three primary kingdoms would have come into being simultaneously, while in the other cases they would have been generated by two successive separations that would represent *the first two dichotomies of the history of life*.

In order to make a choice among the four possibilities of Figure 6.1, we need to compare two or more molecules that were present before the dichotomies and were transmitted to all three kingdoms, and it turned out that such a comparison could in fact be made. The molecules in question were Tu and G elongation factors (Iwabe *et al.*, 1989), and some ATPase subunits (Gogarten *et al.*, 1989), and the experimental results proved that the real phylogenetic tree is that described by Figure 6.1D.

We conclude that there have been two distinct dichotomies at the dawn of the history of life, one that gave rise to eubacteria, and another that separated archaebacteria from the eukaryotic line. The common ancestor that gave rise to the primary kingdoms did not necessarily have the characteristics of Woese's progenote, and for this reason today is referred to as LCA (last common ancestor). Such a term implies of course the existence of a FCA (first common ancestor), and has been adopted because it is reasonable to assume that the very first cells which appeared on Earth went through a number of changes before separating into the primary kingdoms. We conclude that after the origin of life there have been three distinct stages of cellular evolution (Figure 6.2):

(1) The transition from first to last common ancestor.

(2) The dichotomy that gave rise to eubacteria.

(3) The dichotomy that gave rise to archaebacteria.

At the moment, there is no agreement about the names of the other

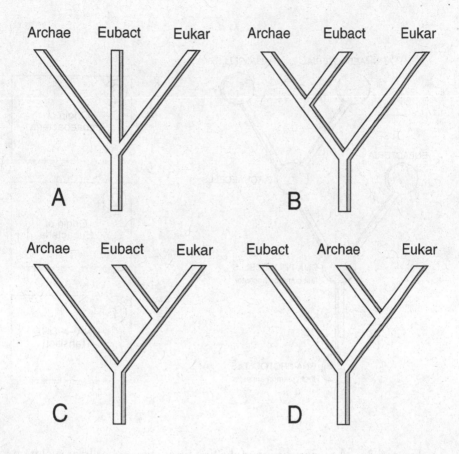

Archae = Archaebacteria

Eubact = Eubacteria

Eukar = Eukaryotes

Figure 6.1 The species correlation coefficients obtained by Carl Woese from ribosomal RNAs revealed the existence of three primary kingdoms at the base of life's evolutionary tree, but are compatible with four different phylogenetic relationships among the primary kingdoms.

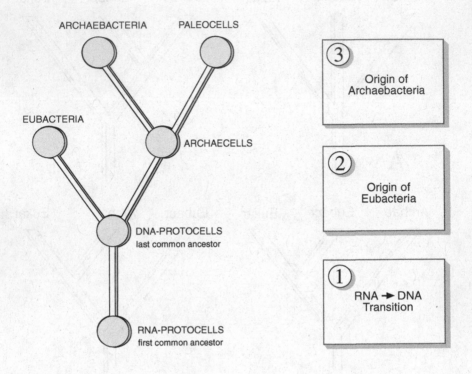

Figure 6.2   A reconstruction of the first three stages of cellular evolution.

cell types that were produced in these dichotomies, and the names in Figure 6.2 (*archaecells* and *paleocells*) represent only my own particular choice. Names apart, however, the phylogenetic tree of Figure 6.2 is a summary of all the experimental information that we have on the origin of the primary kingdoms, and can be regarded as a schematic but faithful description of the first steps of cellular evolution.

The last common ancestor

The experimental data that prove the existence of the three primary kingdoms do not tell us much about the last common ancestor, but we can still say that such a progenitor must have existed, because all cells of the three kingdoms have the same genetic code, the same metabolic currency based on ATP, and roughly 50% of bacterial genes have homologues in eukaryotes.

As for the first living cells (the first common ancestor) we know even less, but again we are not completely in the dark. The evidence that we do have tells us that they came from the ribotype world, and therefore their genomes were made almost completely of RNAs. This means that during the transition from first to last common ancestor, the cells substituted RNA with DNA in their genes, probably by using enzymes that were very similar to reverse transcriptases. Traces of this substitution, in fact, seem to have survived, because many modern enzymes that produce DNA (the DNA polymerases) are still capable of functioning as reverse transcriptases (Poole *et al.*, 1998).

The reason for adopting DNA genomes was probably the fact that DNA is more stable than RNA, and was therefore a more suitable material for heredity, but there could also have been another reason. RNA's linear molecules can be used both as genes and messengers, and in the ribotype world these two roles were performed by the same molecules, which could have created some confusion. The necessity to distinguish between genes and messengers could well have been a good reason for subtituting the RNA genes, and in this case it is not surprising that the choice fell on DNA, because this molecule is easily obtained from RNA and conveys exactly the same information.

In addition to changing the genome's nucleic acids, it is possible that other modifications took place during the evolution from first to last common ancestor, but for the moment we know virtually nothing about these developments. The characteristics of the last common ancestor are therefore highly hypothetical, and yet many have already decided that they were bacterial features. Such a conclusion has dominated evolutionary biology from Haeckel onwards, and is still very popular, because it is based on an apparently incontrovertible concept: the idea that bacteria are primitive because they are the simplest known cells.

In reality, the (relative) simplicity of bacteria can be best explained by the idea that it was the result of a *streamlining* process, just as modern computers have been obtained by a simplification of progenitors that were bulkier, heavier and slower. The properties that we should attribute to the common ancestor, in other words, are not necessarily the most *simple*, but rather the most *primitive*, and with this criterion we can obtain at least three interesting conclusions.

(1) *The link between transcription and translation*

One of the main bacterial features is the fact that DNA transcription is immediately followed by translation, to the extent that in most cases protein synthesis starts on primary transcripts that are still attached to DNA. The result is that there is neither the time nor the space for a modification of the transcripts. In the ribotype world, on the other hand, the first nucleic acids were mostly random molecules, and the first systems were necessarily full of statistical RNAs. It is likely therefore that some kind of screening had to be made before protein synthesis, which means that primitive translation was taking place some time after primitive transcription. A system that contains both useful and useless RNAs is more primitive than a system in which all RNAs are useful, and so it is likely that in the common ancestor transcription was separated from translation.

(2) *The regulation of protein synthesis*

Bacteria can rapidly adapt to changing environmental conditions because their control of protein synthesis is based on *unstable*, or *short-lived*, messengers. Such a fast-reacting system, however, could hardly be primitive, because there is no necessary link between external stimuli and internal messengers, and cells could have learned to build one only after a long evolutionary process. It is likely therefore that the first cells had stable (or long-lived) messengers, because only these molecules could have allowed the evolution of primitive control systems, while unstable messengers would have required advanced forms of regulation.

(3) *The genome organization*

The bacterial genome consists of a single circular DNA molecule, where all genes carry real information and are arranged one after the other without interruptions. Such an organisation is surely very efficient, but precisely for this reason it could not have been present at the beginning. A genome which consists of many, open-ended, DNA molecules is definitely more primitive that a bacterial one, and it is also more likely that the first chromosomes did not contain only nucleic acids, but other molecules as well.

As we can see, the features that one can reasonably attribute to the common ancestor are not bacterial ones, but the very features that later, in a more complex form, will be found in eukaryotes. The separation between transcription and translation, the use of stable messengers, and a genome organised in linear chromosomes are all typical eukaryotic characteristics, and yet they are also intrinsically primitive features.

The last common ancestor did not have the impressive structures that we usually associate with eukaryotes – it did not have a nucleus, a cytoskeleton, mitochondria, chloroplasts, mitosis, meiosis or sexuality – and yet it did already have the basic features that deep down characterise the eukaryotic cell. Despite the lack of a nucleus, in short, the last common ancestor was not a bacterium, because it did not have the *functional* features that are specific of bacteria.

The origins of bacteria

The primitive oceans had the consistency of a diluted broth, and it is likely that their organic molecules were used by the first cells as nutrients. Even such a large food store, however, was inevitably destined to become extinguished, and this created the conditions for the appearance of two very different survival strategies. Some cells adapted their metabolism to smaller and smaller starting molecules, and eventually learned to perform all metabolic reactions from inorganic compounds. In this way they ceased to be *consumers*, and became *producers* of organic matter (and when this happened, the risk that life could become extinct by lack of food finally disappeared).

Other cells continued to feed on organic matter, but in order to do so they were forced to use increasingly big compounds. A potentially important source of food was provided by the bodies of other cells, especially dead ones, and the consumers learned to develop structures that enabled them to ingest bigger and bigger pieces of organic matter. Such a property, that later became true *phagocytosis*, required a plasma membrane that was able to assume the form and the movements of a "mouth", and to do this the cells had to be capable of changing their shape from within, by using molecular structures that eventually evolved into *cytoskeletons*.

There were, in other words, two possible reactions to the alimentary crisis produced by dwindling organic sources, and the descendants of the first cells explored both of them, thus giving rise to two divergent evolutionary lines. The interesting thing is that we can reach this conclusion even by a totally different route.

Let us put aside for a moment the food problem, and let us assume that the descendants of the first cells had to cope with a transformation of the ancient potassium seas into sodium-dominated oceans. In this case the threat was represented by the osmotic effects of sodium, and cells could counterbalance it in two ways: either by developing a rigid external wall that prevented the cell membrane from swelling, or by building an internal net that could bind the plasma membrane from within. Each solution, however, could be realised only under certain conditions.

A rigid external wall would let only small molecules through, and could be adopted only by cells which managed to survive with small metabolites. An internal net of filaments that tied the plasma membrane was a more complex solution, but those cells that were incapable of surviving with small metabolites had no other choice. They could not adopt the external wall solution, and could counteract the sodium osmotic pressure only by anchoring the cell membrane from within, with a network of filaments.

As we can see, the evolution of two divergent cell lines could have been provoked either by a food crisis or by an osmotic danger, or even by both factors. We conclude therefore that the first dichotomy of cellular evolution (Figure 6.2) was essentially a separation between *producers* and *consumers* of organic molecules, and it wouldn't be surprising if such a divide had more than one cause, because any type of metabolism is a solution to a plurality of problems.

The second dichotomy, i.e. the divide that gave rise to archaebacteria, appears to be simpler to explain, at least at first sight. Archaebacteria are micro-organisms which have unusual properties, and have been called *extremophiles* because they are all adapted to extreme conditions. *Thermophiles* and *hyperthermophiles*, for example, grow at temperatures between 80 °C and 120 °C, especially in oceanic and terrestrial hydrothermal vents. *Psychrophiles* live in extremely cold environments, between 0 °C and 4 °C, and even stop reproducing when temperatures rise above 12 °C. *Halophiles* grow in highly salty niches, such as salt-evaporation basins. As for pH, there are two different types of extremophiles: *basophiles* prosper in habitats, such as soda lakes, whose pH is greater than 9, while *acidophiles* colonise areas with pH between 1 and 5, like sulphur vents and peat bogs.

The extremophiles are clearly a case of adaptation to exceptional environments, and this gives the impression that they are rather easy to account for, but we should not forget that micro-organisms had another alternative. It is true that their ancestors could have adapted to extreme environments, but it is also true that they could have avoided those places altogether. Some could have been trapped, and been left with no choice but to adapt, but it is difficult to think that this is what happened in all cases. It is more likely that primitive micro-organisms colonised extreme environments not for lack of choices, but because they were physiologically *predisposed to make experiments*. This hypothesis suggests that primitive cells could live equally well in different environments because they had not yet developed a rigid metabolism, and had a great metabolic plasticity.

Whatever happened, in any case, there have been at least two evolutionary experiments which ended with the "discovery" of the bacterial cell. We know that the two experiments have been independent, because the gulf that divides eubacteria from archaebacteria is simply enormous, but we also know that in both cases they ended up by producing bacteria, and this is highly instructive. It tells us that the bacterial cell was not a *starting-point*, but an *end result*, and it also tells us that there were different ways of achieving that result. The bacterial cell becomes in this way almost a "logical" solution that was discovered many times over, not an isolated accident that was produced by an extraordinary piece of luck at the beginning of life.

The cytoskeleton

A cytoskeleton is absolutely essential for typical eukaryotic processes such as phagocytosis, mitosis, meiotic sexuality, ameboid movement, nuclear assembly and the chromosomes' three-dimensional organisation, i.e. for all those features that make eukaryotic cells so radically different from bacteria. It is not surprising therefore that a large consensus exists on the idea that the origin of the cytoskeleton was probably the most important invention for the development of the eukaryotic cell (Cavalier-Smith, 1987). The stages of eukaryotes' evolution are still shrouded in mystery, but it seems reasonable to assume that the first cytoskeletons were developed either to favour the movements of phagocytosis or to protect the cells from osmotic damage by sodium.

The evolution of the other stages is more difficult to reconstruct, because the cytoskeleton is in reality an integrated system of three cytoskeletons made of specific molecular fibres (microfilaments, microtubules and intermediate filaments) which give complementary contributions to the three-dimensional form of the cell and to its mobility. Perhaps it would be more accurate to say that the system behaves both as a *cytoskeleton* and as a *cytomusculature* (as suggested by Alberts *et al.*, 1989), but even these terms are not entirely appropriate, because they do not convey the idea that the cytoskeleton and the cytomuscles are in a continuous state of assembly and disassembly, and can assume very different three-dimensional configurations. It is as though an animal could sometimes have the skeleton of a bird and at other times the skeleton of a snake, or as though the same muscles could move as a hand or as a heart according to circumstances.

The driving force of the cytosleton is a very unusual mechanism which biologists have decided to call *dynamic instability*: the cytoskeletal filaments – especially microtubules and microfilaments – are in a state of continuous flux where monomers are added to one end and taken away at the other, and the filament is growing or shortening according to which end is having the fastest run. But what is really most surprising is that all this requires energy, which means that the cell is investing enormous amounts of energy not in building a structure but *in making it unstable!*

There is a striking contrast between the sober efficiency of bacteria and the continuous energy "waste" of eukaryotes, and this has made many biologists conclude that the eukaryotic cell is a baroque creature, a lavish squanderer, a true miniature rainforest where everything is luxuriant, overflowing and extravagant. Such a conclusion, however, takes for granted that dynamic instability is an *optional*, not a *necessary* mechanism. But necessary for what? What is it that can justify such a profusion of energy?

In order to understand the logic of dynamic instability, we need to keep in mind that cytoskeletal filaments are unstable only when their ends are not attached to particular molecules that have the ability to anchor them. Every microtubule, for example, starts from an organising centre (the *centrosome*), and the extremity which is attached to this structure is perfectly stable, whereas the other extremity can grow longer or shorter, and becomes stable only when it encounters an anchoring molecule in the cytoplasm. If such an anchor is not found, the whole microtubule is rapidly dismantled and another is launched in another direction, thus allowing the cytoskeleton to explore all the cytoplasm's space in a short time.

A classic example of this strategy is offered by mitosis. In this case it is imperative that microtubules become attached to the centromeres, so that the chromosomes can be transported to opposite ends of the splindle, but centromes are extremely small and their distribution in space is virtually random. Looking for centromeres is literally like looking for a needle in a haystack, and yet the exploratory mechanism of dynamic instability always finds them, and always manages to find them in a brief span of time.

Now the logic is beginning to emerge. Dynamic instability is a mechanism that allows the cytoskeleton to build structures with an *exploratory strategy*, and the power of this strategy can be evaluated by considering how many different structures it can give rise to. The answer is astonishing: the number of different structures that cytoskeletons can create depends only upon the choice of anchoring molecules, and is therefore potentially *unlimited*.

It is the anchoring molecules (that strangely enough biologists call *accessory proteins*) that determine the form that cells have in space and the movements that they perform. The best proof of this enormous versatility is the fact that the cytoskeleton was invented by unicellular eukaryotes, but later was exploited by metazoa to build completely new structures such as the axons of neurons, the myofibrils of muscles, the mobile mouths of macrophages, the tentacles of killer lymphocytes and countless other specialisations.

We conclude that dynamic instability is a means of creating an endless stream of cell types with only one common structure and with the choice of a few anchoring molecules. But this is possible only because there is *no necessary relationship* between the common structure of the cytoskeleton and the cellular structures that the cytoskeleton is working on. The anchoring molecules (or accessory proteins) are true *adaptors* that perform two independent recognition processes: microtubules on one side and different cellular structures on the other side. The resulting correspondence is based therefore on *arbitrary* rules, on true natural conventions that we can refer to as the *cytoskeleton codes*.

The compartments

The plasma membrane of bacteria can be compared to a cellular "skin" because it contains structures that synthesise its molecules *in situ*, just as a true skin contains the cells that continually renew it. In eukaryotes, instead, the plasma membrane is produced by a completely different mechanism. The membrane's "pieces" are made in the cell's interior as vesicles that move towards the surface, and here become incorporated into the existing membrane, while other vesicles detach themselves from the plasma membrane and move towards the interior to be recycled. In eukaryotes, in other words, the plasma membrane is the continuously changing result of two opposite flows of vesicles, and its integrity is due to the perpetual motion of these ascending and descending currents.

Once again, it seems that eukaryotes invented an extremely expensive mechanism for a purpose that could have been achieved much more simply, as bacteria clearly prove. This has been regarded as yet another example of eukaryotic extravagance, and some have even suggested that it is useless to look for an engineering logic in eukaryotes, because these are creatures that love exaggeration and waste, not efficiency and economy.

The building mechanism of the eukaryotic membrane, however, can be seen in a totally different light if we regard it not as an isolated case, but as an example of a wider class of phenomena. More precisely, as one of the various mechanisms that eukaryotic cells employ to build their *compartments*. The vesicles that are destined for the plasma membrane, in fact, are produced in the Golgi apparatus together with vesicles which have very different destinations. Some are delivered to lysosomes and others to secretory granules.

The Golgi apparatus is involved in the terminal modification of innumerable molecules which have diverse destinations, and if every molecule had to follow a specific path, the cell simply could not cope with the immensely intricate traffic that would have to be directed. The Golgi apparatus, instead, delivers to their destinations an astonishing number of molecules with only two types of vesicles: one for transporting proteins outside the cell, and the other to its interior. This requires only two destination signals for the vesicles, however large is the number of transported proteins. On top of that, the Golgi apparatus produces a third type of vesicles which do not carry any destination signal, and these are the vesicles that are programmed, *by default*, to reach the plasma membrane. As we can see, the solution is extraordinarily efficient: with a single mechanism and only two types of signals, the cell carries an enormous amount of specific products to their destinations, and also manages to continually renew its plasma membrane.

The Golgi apparatus, however, is a transit place for only a fraction of the proteins which are actually produced by eukaryotic cells. The synthesis of all such proteins invariably begins in the soluble part of the cytoplasm (the *cytosol*), and during this first step they also receive a signal that specifies their geographic destination. The piece of the amino acid chain that emerges first from the ribosome machine – the so-called *peptide leader* – can contain a sequence that the cell interprets as an *export signal to the endoplasmic reticulum*. If such a signal is present, the ribosome binds to the reticulum and delivers the protein into its *lumen*. If the peptide leader does not carry such a signal, the synthesis continues on free ribosomes, and the resulting proteins are shed in the cytosol. Of these, however, only a fraction are destined to remain there, because the amino acid chain can carry, in its interior, one or more signals which specify other destinations. More precisely, there are signals for protein export to the *nucleus*, to *mitochondria*, and to other cell compartments. Proteins, in conclusion, carry with them the signals of their geographic destination, and even the absence of such signals has a meaning, because it implies that the protein is destined to remain in the cytosol.

The crucial point is that there is *no necessary correspondence* between protein signals and geographic destinations. The export-to-the-nucleus signals, for example, could have been used for other compartments, or could have been totally different, just as the names which are given to cities, to airports and to holiday resorts. The existence of eukaryotic compartments, in conclusion, is based on natural conventions, and to these rules of correspondence we can legitimately give the name of *compartment codes*.

Chromosomes

Bacteria have a single chromosome which has a circular form and no stable association with structural proteins, while eukaryotes contain various chromosomes which are open-ended (or linear) molecules, and bind large amounts of structural proteins which fold the DNA thread many times over. The classic *double helix*, discovered by Watson and Crick in 1953, has a width of 2 nm, but in eukaryotes many segments of this filament are folded around groups of eight proteins (called *nucleosomes*) which give to the structure a "beads-on-a-string" appearance. This chromatin string is almost six times thicker than the double helix, and represents the so-called *11-nm fibre*, but *in vivo* it is always folded into spirals of nucleosome groups (called *solenoids*) to form the *30-nm fibre*. When the cell is dividing by mitosis, furthermore, chromosomes are further condensed to give rise first to a *300-nm fibre*, then to a *700-nm fibre* and finally to the full *1400-nm metaphase fibre*.

These foldings enormously reduce chromosome lengths, and allow eukaryotic cells to carry many DNA molecules, each of which is on average a thousand times longer than a typical bacterial chromosome. The eukaryotic cell is clearly capable of carrying many more genes than bacteria, and this suggests that its genome evolved precisely for that reason, but nature, once again, is full of surprises. It is true that eukaryotes have more genes than bacteria, but it is also true that by far the greater part of eukaryotic DNA does not form genes. It is as if eukaryotes had developed to the highest level the strategy of transporting enormous amounts of DNA most of which is apparently useless from a genetic point of view.

This discovery has been one of the greatest surprises of molecular biology, and represents a paradox that has not yet been solved. Eukaryotes have a highly redundant genome, in which only a tiny fraction is transporting genes, and once again it seems that the sole explanation is an incorrigible eukaryotic drive toward lavishness and exaggeration. Today we do not have experimental data that justify such a waste of resources, but perhaps we can still attempt some speculations.

If the first cells had genomes that contained both functional and "junk" RNAs, they also had before them two different evolutionary strategies: one was to get rid of the "junk" RNAs, the other was to keep them. The first strategy led to a generalised streamlining, and was embraced by those cells that eventually became bacteria. The second was adopted by cells that did not leave the "old road", and went on transporting molecules of RNA, and later of DNA, that did not have genetic functions.

These "junk" nucleic acids were physically interspersed between genes and could have had a variety of effects. Some, for example, could have divided genes into clusters, and in each cluster could have kept the genes apart, so that the whole target area was more easily found by signalling molecules. Others could have attached themselves to various cellular structures, thus forming a primitive three-dimensional scaffolding for the entire genome. We still know far too little about the three-dimensional features of eukaryotic genomes, but we do know, for example, that centromeric regions have essentially a *mechanical* role, not a genetic one, and the same could be true for other chromosomal regions.

It may be useful, furthermore, to keep in mind an interesting but still mysterious result obtained by Mantegna and collegues in 1994. These authors applied to genomes two renowned formulas (Zipf's law which is valid for all known languages, and Shannon's redundancy rule), and found that the non-coding regions have statistical properties more similar to those of natural languages (smaller entropy and greater redundancy) than the coding regions. This suggests that the non-coding regions of the eukaryotic genome are probably conveying information, but are doing so in a *different language*, about which we know nothing.

As we can see, it is far too early to draw reliable conclusions about eukaryotic genomes, but once again we can appreciate how important it is to distinguish between *simple* and *primitive* properties. The streamlining strategy did succeed in producing cells that combined maximum efficiency and maximum simplicity, but when all that was erasable was in fact erased, it become impossible to go further, and bacterial evolution reached a sort of stationary state. Other cells, instead, went on with primitive cargoes and "junk" molecules, and their evolution remained open, free, creative and unpredictable.

The seven kingdoms

Molecular data did allow us to reconstruct the primordial dichotomies that gave rise to eubacteria and archaebacteria (Figures 6.1 and 6.2), but so far have not revealed much about the other stages of cellular evolution. The eukaryotic cell is such a complex labyrinth that it is very hard to understand how its pieces were put together. It is even possible that its structures were so thoroughly mixed that historical traces have been lost forever. It is also true, however, that molecular phylogeny is still a young science, and we cannot exclude the possibility that one day it may have some surprises in store for us.

For the time being, we can reconstruct only a few great events of cellular evolution on the basis of clearly visible morphological changes. More precisely, after the first three stages of cellular evolution we can clearly recognize only four other major stages (Figure 6.3):
(a) The appearance of a nucleus in some paleocells gave rise to *paleokaryotes* (step 4).
(b) Some paleokaryotes acquired mitochondria and chloroplasts by symbiosis and generated the first *protists* (step 5).
(c) Some protists went on developing new eukaryotic characters, and a few even started experimenting with *multicellularity* (step 6).
(d) Some experiments in multicellularity had success and gave rise to the first progenitors of *plants, fungi and animals* (step 7).

The sequence of these steps gives us a fairly accurate reconstruction, not only because any step needs the previous one *as a biological prerequisite*, but also because the whole sequence is in good agreement with the data from paleontology. We conclude therefore that there have been at least seven major steps in cellular evolution, and that such steps gave rise to the highest taxonomic groups that we call *kingdoms* (Figure 6.3).

It must be stressed that today there still is no general agreement not only about the names, but also about the number of the kingdoms. Some, for example, do not accept that Paleokaryota (cells that did not acquire organelles by symbiosis) represent a kingdom. Cavalier-Smith, on the other hand, has given the name of *Archezoa* to these primitive micro-organisms, but has ranked

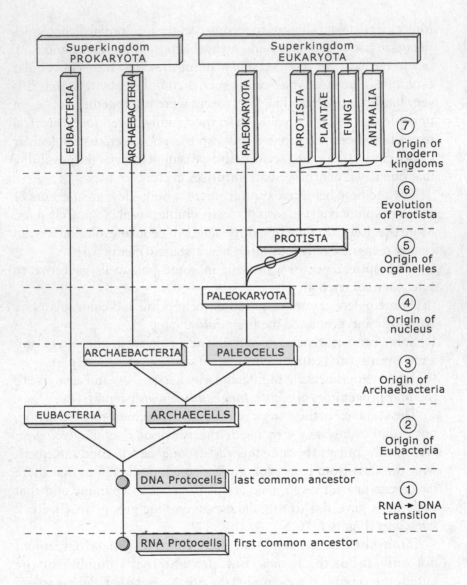

Figure 6.3   Schematic evolution of the seven kingdoms of life in seven stages. The white rectangles indicate kingdoms that did leave modern descendants, while the grey frames are for extinct kingdoms.

them as a *subkingdom* (1987), while Maynard Smith and Szathmáry (1995) have regarded them as a *superkingdom*.

The scheme of Figure 6.3 represents what seems to me the most reasonable solution: it does accept that some Paleokaryota left modern descendants – and this qualifies them as a true kingdom – but also accepts that some Archezoa could have lost organelles secondarily, and are not therefore true Paleokaryota.

The most important information in Figure 6.3, however, does not concern the number or the names of the taxonomic kingdoms, but the distinction between *extinct kingdoms* (grey boxes) and *surviving kingdoms* (white boxes). The data provided by molecular phylogenies are all obtained from *living organisms*, i.e. from representatives of surviving kingdoms, and clearly they will never give us direct information about extinct kingdoms. If we use only molecular phylogenies, in other words, we are bound to conclude that *extinct kingdoms did not exist*, and that the first living cells were eubacteria.

In order to have a less biased reconstruction of the early history of life, the data from molecular phylogenies must be integrated by theoretical considerations. Those data are absolutely necessary – no doubt about that – but we should not forget that they are not sufficient. The history of life on Earth does not coincide with the history of the molecules that have survived, as if extinct kingdoms never existed. Molecules, therefore, take us back in time but only to a certain point. Beyond that, only geology and our theories can make the journey.

Three thousand million years

Let us imagine that an extraterrestrial civilization wanted to study our planet's life and decided to send a spaceship on Earth once every million years. For at least 2000 times, the answer would have always been the same: *"The dry lands are completely sterile, and in the seas there are only colonies of bacteria."* After that, the dispatches would have been slightly different: *"Now there are small amounts of oxygen in the atmosphere, and the seas are also inhabited by bigger cells which have a nucleus."* This verdict would have been repeated for about another 1000 times, and the extraterrestrials would have had every right to conclude that life's evolution on Earth was exasperatingly slow. Within a few other missions, however, everything would have changed: *"Life has really exploded on Earth, and many multicellular organisms, of weird shapes and sizes, are not only swimming in the oceans but also running on land and flying in the skies."*

This is precisely how we reconstruct (with far fewer samples) our planet's past. The first cells appeared on Earth more than 3.5 billion years ago, and the first animals arrived just over 500 million years ago. For 3 billion years, in other words, the Earth was inhabited only by micro-organisms. Three billion years are simply inconceivable to our mind. We can try alternative expressions such as *three thousand million years* or *three million millennia*, but none of these wordings can give us even a feeble idea of the immensity of that expanse of time.

The only thing about which there seems to be a general agreement is that cellular evolution was incredibly long and incredibly slow: it appears that very little happened in the first 3000 million years, and that the real story of life started in earnest only at the end of that enormously boring aeon, with a spectacular explosion of creativity. Things, of course, could have gone precisely that way, but let us try to look back from a different point of view.

An army, a crowd, a nation, or other human societies, can be very complex structures, but none of them is as complex as a single individual human being. And the same is true for all animal societies: a beehive or an ant-hill, for example, is a far simpler system than any individual insect. And if it is true that individuals are more complex than their societies, then it could be that single eukaryotic cells are more complex than societies of eukaryotic cells, i.e. multicellular organisms. And this in turn would explain why unicellular evolution had to be so much longer than multicellular evolution.

This is an attractive hypothesis, but it does have a weak point. It is true that societies such as armies, crowds, beehives and ant-hills, are simpler than individuals, but this happens because they are made of individuals which are *physically separated.* The same does not apply to animals and plants, where individual cells gave up their physical freedom and became integrated units of a greater individual. A proof of this is that the complexity of the nervous system, or of the immune system, is certainly not inferior to the complexity of the eukaryotic cell. We seem bound to conclude, therefore, that animal evolution created complex structures, like our nervous system, in 500 million years, whereas cellular evolution needed some 3000 million years to generate a modern eukaryotic cell. Cellular evolution, in other words, does appear to have been objectively slower that animal evolution.

Up until a few years ago, such a conclusion was practically inevitable, but today we are not so sure. The evolution of the nervous system and the evolution of the eukaryotic cell would really be comparable *only if their starting-points had been equivalent*, and this is not true. In embryonic develoment, for example, nerve cells use an exploratory strategy which is based on the dynamic instability of the cytoskeleton, but did not have to invent that strategy from scratch, because eukaryotic cells already provided it. And the same is true for virtually all other major metazoan inventions. The endocrine system, for example, exploits the same signalling mechanisms that free-living eukaryotic cells had already developed. The differential expression of genes – which is at the very heart of embryonic differentiation – had been invented long before by unicellular eukaryotes which learned to use different parts of their genomes in different phases of the cell cycle. And so on and so on.

It is true, in conclusion, that animal evolution was fast, but it is also true that *countless crucial inventions had already been made before*. And this throws a completely different light on those first 3000 million years. It is certainly true that that was not a time of great morphological changes, but what went on inside the cells is a totally different story: perhaps it was precisely there that the greatest part of creation took place.

7

THE CAMBRIAN EXPLOSION

The appearance of all animal phyla in a narrow geological stratum at the base of the Cambrian has been one of the greatest discoveries of all times, but has also been, and still is, one of the greatest unsolved problems of biology. What has to be explained is primarily the origin of all animal phyla in a geologically brief period of time, but there are other two closely related problems which are also waiting for an answer. One is the fact that the animal body plans have been strongly conserved after the Cambrian explosion, but not before. The other problem comes from embryology. The body plan is built during development and becomes visible at the phylotypic stage, the brief period in which the embryos of all the species of a phylum have the greatest degree of similarity. Before and after that stage, the embryos of different species become increasingly diverse, but in a very puzzling way. Before the phylotypic stage the differences are decreasing, whereas after that stage they are steadily increasing. The Cambrian explosion, in short, gives us three major problems: the origin of the animal phyla, the conservation of the body plan and the conservation of the phylotypic stage. In this chapter it is shown that if animals are described as *idealised multicellular structures* which are reconstructed from incomplete information, all three problems can be solved. Needless to say, it is a solution of an *idealised Cambrian explosion* obtained with *idealised organisms*, but the principles involved should not be dismissed lightly. They are very general and may well apply to all cases.

The fossil record

At the end of the eighteenth century, William Smith, an English
engineer who was engaged in canal building, discovered an empirical
rule for comparing the rocks of different geographical areas. The idea
was to identify the sedimentary rocks by their fossils, because Smith
had noticed that each stratum contains fossils that are never found in
higher or in lower strata. Hence the idea that if two rocks have the
same fossils, they also have the same geological age, even if one is at
the bottom of a valley and the other on the top of a mountain. The
fossils became in this way the key for reconstructing the past
movements of the Earth's crust, and Smith used them to draw the
first geological map of the United Kingdom.

Smith's discovery, however, had implications that went far beyond
geology. The fact that the fossils of a stratum do not appear in all
other strata means that the organisms which lived in that geological
age were different from those that lived in all other ages. It means
that life on Earth has gone through a long history of changes, and
that sedimentary rocks are still keeping a record of that history.

But how accurate is the fossil record? What we see in it is all that
remains of the ancient inhabitants of the Earth, but the remnants
could be imperfect, and even deceitful, documents. The sedimentary
rocks contain, sometimes within a few centimeters, materials that
were deposited to the bottom of ancient seas for millions of years,
and their sharply discontinuous structure seems to suggest that
organisms appeared suddenly on the face of the Earth only to
disappear, some time later, with the same abruptness. Can this be a
faithful testimony of what really happened? The fossils interpretation
problem was forcefully brought to general attention when the greatest
of all discontinuities was discovered, the so-called *Cambrian
explosion*. In the 1830s, Roderick Murchison found the geological
stratum that contained the very first visible fossils of the Earth's
history. Whilst all lower strata were, to the naked eye, completely
devoid of fossils, in the Cambrian one could see the fossilised remains
of creatures that unmistakably exhibited the sophisticated structures
of highly developed Metazoa.

The Cambrian animals were totally different from modern organisms, but Murchison did not see in this a sign of evolution. The most telling fact, to him, was the suddenness of their appearance, and he concluded that the abrupt arrival of complex animals could only be explained by an act of creation. *"The first signs of living things,"* he wrote in 1854, *"announcing as they do a high complexity of organization, entirely exclude the hypothesis of a transmutation from lower to higher grades of being. The first fiat of Creation which went forth, doubtlessly ensured the perfect adaptation of animals to the surrounding media."*

Darwin could reply to this argument only by invoking the imperfection of the fossil record, and almost a century had to pass before George Simpson could point out that a geologically sudden event has a time span of a few million years, and therefore is not at all sudden from a biological point of view. Today, Simpson's argument has been largely confirmed by the experimental data, but it should not be forgotten that the Cambrian explosion is still waiting for an explanation. And what has to be explained is not only its physical time span but also, and above all, its biological mechanism.

The experimental data

It has been very difficult to obtain reliable data about the Cambrian, and many conflicting proposals have been made both on the dating of its geological strata and on its division into subperiods. For a long time, for example, the beginning of the Cambrian has been associated with the appearance of trilobites (its most characteristic animals), but then it was found that these arthropods were preceded by the so-called *small shelly fossils*, minute animals with tiny shells which appeared – again "suddenly" – in a lower Cambrian stratum which was called *Tommotian* (from the name of a Russian site). For a number of years the Tommotian marked the beginning of the Cambrian, but even this conclusion has turned out to be unsatisfactory. The rocks immediately below it, in fact, can also be included in the Cambrian, but are practically devoid of small shelly fossils, and show instead a great number of *trace fossils*, that is to say tunnels and burrows that were undoubtedly excavated by small animals. Trace fossils do not allow us to reconstruct the animals that made them, but we can still divide those animals into groups, and even into distinct taxa. The most characteristic taxon was named *Phycoides pedum*, and eventually an international convention established that the appearance of *Phycoides* trace fossils marks the beginning of the Cambrian.

The first stage of the Lower Cambrian is known today as *Manykayan* (trace fossils), after which come the *Tommotian* (small shelly fossils), the *Atdabanian* (trilobites), and the *Botomian*, while the other two stages of the period are still known with the generic names of Middle and Upper Cambrian (Figure 7.1).

In the last twenty years, in conclusion, paleontology has discovered that in Cambrian times there have been not one but three different "explosions" of animal life: one documented by trace fossils, a second which left behind small shelly fossils, and finally the classical explosion that was dominated by trilobites. It must also be added that Cambrian life was preceded by the so-called *Ediacara fauna*, a vast assembly of soft-bodied animals (almost all with radial symmetry). Many scholars now regard them as a failed evolutionary experiment, while others believe that they may have left modified descendants (Figure 7.1).

Another problem which has been fraught with technical difficulties is the dating of Cambrian rocks with radioactive methods. From 1993 onwards, however, the results have been fairly trustworthy, and today we regard them as almost definitive, in the sense that great variations are no longer expected from future measurements (Gould, 1989; Conway Morris, 1993; Fortey, 1998). On the basis of what has reasonably been established, we can be confident that the Cambrian period started around 545 million years ago, and lasted some 40 million years. Another important conclusion is that the Manykayan went on for some 10 million years, while the Tommotian and the Atdabanian were both in the range of 2 to 5 million years. The explosion of small shelly fossils and the classical explosion of trilobites, in other words, did not exceed 5 million years, and today this does look like the maximum time span of those great transformations. By far the most important discovery, however, has come from biological studies, and more precisely from the comparative anatomy of the fossils. This has

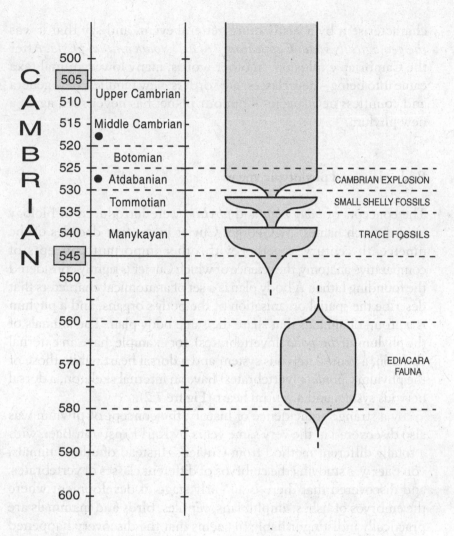

Figure 7.1   The three Cambrian "explosions" of animal life, and the probable temporal extensions (5-10 million years) of their geological strata.

proved that Cambrian animals invented the body plans of all the animals that have appeared on Earth ever since.

The Cambrian explosion was traditionally defined as the appearance of the first skeleton-bearing Metazoa, but now we can characterise it by a vastly more general event, and say that it was *the geologically sudden appearance of all known animal phyla*. After the Cambrian explosion, in other words, many lower animal taxa came into being – new classes, new orders, new families, new genera and countless new species – but our planet has never seen again a new phylum.

Body plans and phylotypic stages

The concepts of *body plan* and *phylum* were introduced in biology (with French names) by Georges Cuvier in the first decades of the nineteenth century, together with other important concepts of comparative anatomy, the science of which Cuvier is rightly considered the founding father. A body plan is a set of anatomical characters that describe the spatial organisation of the body's organs, and a phylum is a group of animals that share the same body plan. The animals of the phylum *Arthropoda* (invertebrates), for example, have an external skeleton, a ventral nervous system and a dorsal heart, while those of the phylum *Chordata* (vertebrates) have an internal skeleton, a dorsal nervous system and a ventral heart (Figure 7.2).

By a strange coincidence of history, the concept of phylum was also discovered in the very same years by Karl Ernst von Baer, with a totally different method from Cuvier's. Instead of adult animals, von Baer was studying the embryos of different classes of vertebrates, and discovered that there is an early stage of development where the embryos of fishes, amphibians, reptiles, birds and mammals are practically indistinguishable (it seems that the discovery happened because one evening von Baer forgot to label the bottles of his samples, and the next day could no longer recognise the class of the embryos).

Later on, von Baer found that that phenomenon has a general validity. All animals can be divided into major groups – today called *phyla* – which are characterised by the fact that the animals of one group have a stage of embryonic development where all their embryos are strikingly similar. This stage of maximum morphological similarity

CHORDATES   ARTHROPODS

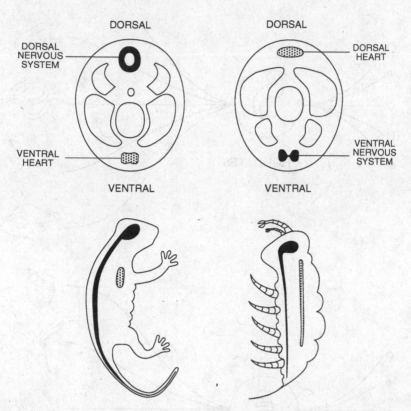

Figure 7.2   The body plan of chordates (deuterostomes) consists of an internal skeleton, ventral heart and dorsal nervous system, while the *Bauplan* of arthopods (protostomes) is characterised by an external skeleton, dorsal heart and ventral nervous system.

is known today as the *phylotypic stage*, precisely because it is a defining characteristic of each phylum (Figure 7.3).

The link between the two discoveries is the fact that the phyla defined by Cuvier, on the basis of the adult body plans, exactly correspond to the phyla defined by von Baer on the basis of the embryos' phylotypic stages, a convergence that turns out to have a

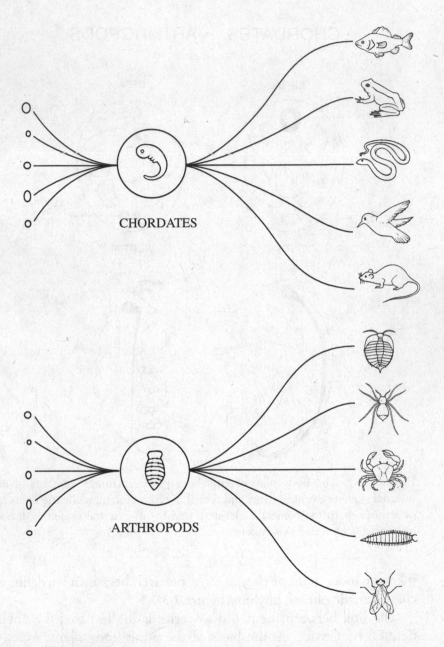

Figure 7.3   The phylotypic stage in chordates (*faringula*) and in arthropods (*segmented germ band*).

deep biological meaning. It is due to the fact that the body plan is
built during embryonic development, and at the phylotypic stage is
already showing all the defining characters of the adult body plan.

This has evolutionary consequences of the greatest importance,
because it means that the historical change of body plans must have
taken place in embryonic life, and more precisely in the stages that
precede the phylotypic stage. Since these stages are known, it is
possible to make a diagram that represents an embryological sequence
of the steps of metazoan evolution (Figure 7.4). Even if an
embryological scheme is not a *historical* one, the diagram does suggest

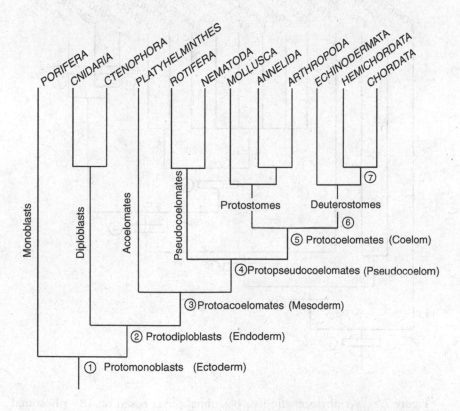

Figure 7.4　A phylogenetic tree of animal phyla based on the stages of
embryonic development. The most significant innovations are the
appearance of (1) ectoderm, (2) endoderm, (3) mesoderm, (4) pseudocoelom,
(5) coelom, (6) protostome–deuterostome dichotomy and (7) dorsal chord.

that Metazoa very probably derived from a common ancestor, and gives us an overall view of the results that were achieved during the Cambrian explosion. The analysis of some RNAs, furthermore, has made it possible to build molecular trees that confirm the monophyletic origin of Metazoa, and which have many similarities with the embryological tree (Figure 7.5). The results of molecular phylogenies are not yet definitive, but the information in Figures 7.4 and 7.5 is

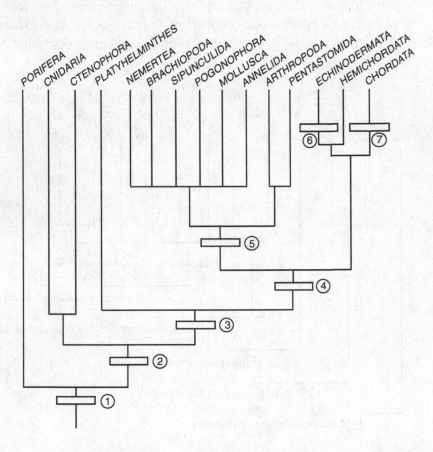

Figure 7.5   A phylogenetic tree of animal phyla based on 18S ribosomal DNA from several laboratories (Raff, 1996). The most significant innovations are the appearance of (1) multicellularity, (2) tissues, (3) the anterior–posterior axis and central nervous system, (5) metameric segmentation, (6) pentameral symmetry, (7) neural crest and amplification of Hox clusters.

fairly illuminating, and it is reasonable to expect that, one day, we will have a detailed molecular reconstruction of the Cambrian explosion. There are, however, two phenomena that remain theoretically difficult to explain: (1) the conservation of Cambrian body plans in all successive periods of evolution, and (2) the conservation of the phylotypic stage in the embryonic development of all animals.

The traditional explanations

So far, three types of hypotheses have been made on the conservation of body plans.

(1) *The environmental explanation*
No new body plan has been invented after the Cambrian explosion because all ecological niches were already occupied.
This is a very weak suggestion, because a high number of new ecological niches did become available after the Cambrian, and not just once but many times over. When animals invaded the land, for example, they had at their disposal absolutely virgin territories for hundreds of millions of years. And as for the sea, species have been literally decimated various times by great mass extinctions, which certainly created plenty of opportunies for new experiments in body plans.

(2) *The embryonic constraints explanation*
Body plans impose so many constraints on embryonic development that any novelty would disrupt too many characters and would bring development to an end.
This seems a more reasonable hypothesis, but surely it cannot apply to Cambrian animals. They too had embryonic developments and body plans, and the Cambrian explosion means precisely that those ancestral plans were modified. If the constraints on embryonic development did not prevent the modification of body plans "before" the explosion, why should have prevented it "after" the explosion?

(3) *The laws of form explanation*
Body plans are the expression of "laws of form" which organisms cannot change, just as minerals cannot change their crystallisation rules.

According to this classical explanation – which today has been reproposed with the name of *biological structuralism* – body plans are immutable either because they embody mathematical laws, or because they are shaped by physical forces that organisms cannot change, just as they cannot change gravity and chemical bonds. Even this explanation, however, collapses before the historical fact that body plans did change in the Cambrian, which means that there is nothing immutable about them.

As we can see, none of the hypotheses that have been proposed so far is satisfactory, and this is probably due to the fact that the *conservation* of body plans and the *origin* of body plans are treated as if they were two disjoined problems. In reality, what we need to explain is not the conservation of body plans *per se*, but *from a certain point onwards*. More precisely, the problem consists in understanding why the body plans were modifiable "before" but not "after" the Cambrian explosion.

The Cambrian singularity

Paleontology has shown that the history of life has been full of *adaptive radiations*, processes in which an ancestral taxon gave rise to descendant taxa which diverged by adapting to different environmental conditions. Some adaptive radiations have been explained with the mechanism of *phyletic gradualism*, while others are better described by *punctuated equilibria*, but in all cases they are classical processes of adaptation to the environment by natural selection.

Some scholars have suggested that the Cambrian explosion too was an adaptive radiation, but this idea would put us before an insurmountable difficulty. Both phyletic gradualism and punctuated equilibria require, *as a theoretical necessity*, that the higher the taxon, the higher the number of speciations, and this means that changes at the phylum level must have taken longer times than changes at lower levels. In the case of the Cambrian explosion, however, this theoretical requirement contrasts with the evidence, and all classical mechanisms of adaptive radiations are incapable of explaining the phenomenon.

The temptation to "declassify" the Cambrian explosion, and to assimilate it to processes that can easily be accounted for, is strong, but leads to an unbridgeable contradiction between experimental data and theoretical previsions. The only reasonable conclusion, therefore, is that the Cambrian explosion was not an adaptive radiation, i.e. it was not a simple process of adaptation to the environment, but something very different.

Such a conclusion is directly suggested by the very characteristics of the explosion. All adaptive radiations that came after the Cambrian have never modified the body plans, while the Cambrian explosion was characterised precisely by modifications of those plans. And, in a similar way, no adaptive radiations have ever changed the phylotypic stage of developing embryos, while the Cambrian explosion did precisely that.

The message that nature herself appears to be sending us is that the Cambrian explosion was a rare event in the history of life, comparable perhaps only to the origin of life or to the origin of the mind. And what is so special about these rare episodes of macro-evolution is *the appearance of biological characteristics which have never been changed ever since.* The mechanism that we are looking for, in conclusion, must explain precisely why the Cambrian explosion was so *different* from a normal adaptive radiation, even if this means that we cannot explain it with classical mechanisms.

The stumbling-block

The fact that we have no explanation for the Cambrian explosion does not seem to surprise anybody, and is indeed understandable, because biologists know only too well where the stumbling-block comes from. The body plan is built during embryonic development by a sequence of genetic and epigenetic processes, but so far we have a reasonable knowledge only of the genetic side of the story. About the epigenetic contribution we are still in the dark, and it is this which prevents us from understanding the "logic" of the whole process.

The Cambrian explosion, on the other hand, had its roots precisely in the developing strategies of Cambrian embryos, and as long as the logic of development remains a mystery, we have no chance of understanding what happened. And we have also little chance of understanding the rest of metazoan evolution, because animals are, first of all, what their embryos make of them.

What is probably less well known is that it is not embryologists who are to be blamed for our present ignorance of epigenesis. It is our culture as a whole that carries that responsibility. The real culprit is the fact that we can build a "genetic" machine (a computer), but not an "epigenetic" system, a machine that is capable of doing what any embryo does. This is the great challenge of the future, and not for embryology alone, but for all of science: *how does a system manage to increase its own complexity?*

While physics has built its fortunes on mathematics, biology is still an essentially empirical science, and mathematical models are mainly used for descriptive purposes, not as guiding principles. In this case, however, the problem is not that biologists are sceptical about mathematical models of epigenesis, but the fact that such models do not exist. As a matter of fact, one does exist, but nobody has taken any notice of it, which amounts to almost the same thing.

It is clear that strong antibodies exist against the idea of a mathematical model of epigenesis, but there simply is no alternative. If we want to understand the Cambrian explosion, we must understand not only the genetic but also the epigenetic side of development, and in order to grasp epigenesis we must have a model that explains how a convergent increase of complexity can be achieved. This is the critical point, and we will therefore try to approach the Cambrian explosion with the assistance of the one and only mathematical model of epigenesis that does exist in the literature.

The reconstruction model

In order to build a mathematical model of embryonic develoment, it is necessary to ignore a multitude of secondary features and to concentrate attention on its fundamental property, i.e. on epigenesis. Even this, however, is not enough. As soon as we specify that epigenesis is *a convergent increase of complexity*, we seem unable to go any further, apparently because there is no satisfactory definition of complexity. We realise in this way that naming the essential feature of development does not help: we also need to translate the problem into an algorithm, if we are to have a working model.

Luckily, a solution does exist: we can start from a different formulation of the problem, and say that embryonic develoment is *a reconstruction from incomplete information.* This is equivalent to saying that new structures appear in stages during development, and is therefore another way of expressing the basic idea of epigenesis. With the new formulation, however, we do not have to provide a definition of complexity in order to build a model. We are facing instead a reconstruction problem, and this we know how to deal with. Tomography, for example, is a reconstruction of structures from radiographic projections and gives us a precious guideline, because there are mathematical theorems that tell us how many projections are required to make a complete reconstruction (Chapter 3). This, in turn, allows us to define what a reconstruction from incomplete information is: *it is a reconstruction from a number of projections that is at least one order of magnitude less than the theoretical minimum that is required to make a complete reconstruction.*

As we can see, the problem can be given a precise formulation, but what really counts is that it can also be given a solution. I have demonstrated that structures can indeed be reconstructed by using only 10% of the minimum number of projections (Barbieri, 1974a, 1974b, 1987), and an iterative algorithm which exploits memory matrices. More precisely, a reconstruction from incomplete projections is possible if two conditions are met: (1) if the reconstruction method employs memory matrices where new information appears, and (2) if the reconstruction method employs codes, or conventions, which transfer information from the memory space to the real space.

The reconstruction of structures, in conclusion, gives us a model that translates epigenesis into a workable problem, and tells us that *memories* and *codes* are the key structures for achieving the goal of a convergent increase of complexity.

Multicellular structures

The reconstruction of individual structures can be a model for the development of individual cells, but not for multicellular systems such as embryos. In this case, a model must be capable of performing in parallel a plurality of reconstructions, while taking into account a variety of interactions between individual structures. What must be simulated is an increase of information between initial and final structures, and on this point the reconstruction model is crystal clear. The difference between initial and final information, and therefore the overall increase of complexity in the system, is entirely dependent on the memories which are used in a reconstruction, because it is only in the memory space that new information appears.

In the case of individual structures, a reconstruction matrix can receive new information only from its individual memory matrix, but in the case of multicellular structures there is also another option. Here it is possible to build a "collective" memory matrix, and this allows us to choose between two different reconstruction strategies. We can continue to adopt an individual approach, where each cell gets new information only from its individual memory matrix (Figure 7.6A), but we can also adopt an approach where, *from a certain point onwards*, a cell can also receive new information from the collective memory of the system (Figure 7.6B).

In the first case, a structure is reconstructed with the same strategy from beginning to end, and we can say that the reconstruction is *continuous*, or *one-phased*. In the other case, the reconstruction is *discontinuous*, or *two-phased*, because after an initial period where the collective memory does not yet exist, comes a second phase where cells can also use the information of the newly-formed collective memory. A discontinuous reconstruction produces a greater increase of complexity than a continuous one (because the collective memory adds additional information), but needs of course a more complex

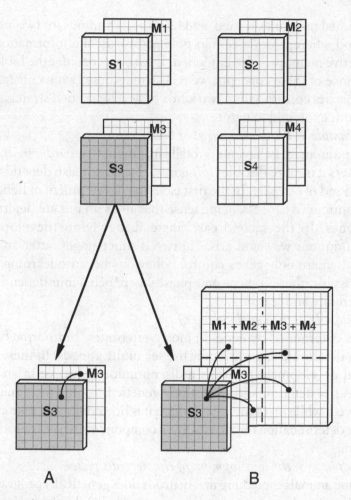

Figure 7.6　The reconstruction of multicellular structures can be performed by using only individual memory matrices (A), or by also exploiting a collective memory matrix (B).

algorithm. The two kinds of reconstruction can therefore be applied to the solution of different problems.

The mathematical model, in short, shows that multicellular structures can be obtained with two different strategies: (1) a continuous reconstruction method where only the information of individual memories is used, and (2) a discontinuous, or two-phased, method where, from a certain point onwards, the information of a collective memory is also exploited. If we now consider the biological relevance of these concepts, we realise that these two reconstruction approaches correspond to two kinds of developmental strategies that are both exploited in nature.

Example 1: the development of behaviour
In some animals, behaviour is totally instinctive (or *hard-wired*), while in others it is more flexible (*soft-wired*), because it also depends upon some kind of *learning*. In the first case, the development of behaviour is a continuous process, in the sense that all its phases are determined by genes. In the second case, instead, behaviour develops in a discontinuous way, and arises in two distinct stages: after an initial period where only genes control behaviour because learning is not yet possible, there comes a new phase where behaviour depends both on heredity and learning.

Example 2: the development of sex
Sex is determined by genes in most vertebrates, but normally what genes determine directly is only the sex of the gonads. In these cases, sexual development is clearly a discontinuous process: in an initial phase, sex determination is entirely genetic but only the gonads are involved, while in a subsequent phase it is hormones that take control of sex determination for all remaining components of the reproductive system.

Example 3: the development of the nervous system
In some animals everything that neurons do is genetically programmed for life, but in many more cases neural development is neatly discontinuous: after an initial phase where the fate of neurons is irrevocably fixed (usually by the time and place of their birth in the neural tube), there comes a second phase where the survival of neurons depends upon the molecules that neural extensions happen to encounter during their exploration of the body.

These examples show that the two strategies proposed by the reconstruction model do indeed correspond to two developmental strategies that have both been exploited by nature. The model is capable therefore of simulating some important characteristics of development, but perhaps it could do better than that: it could even help us to understand something new about the logic of embryonic development.

Biological memories

We have seen that a mathematical model can reconstruct structures from incomplete information only by building "memory matrices" during the process, and this gives us the problem of understanding whether something equivalent occurs during embryonic development. In order to make a parallel between the two cases, we need a definition of "memory" that applies equally well to mathematics and to biology, but this is not a problem. We can say that a memory is *a permanent deposit of information*, because this definition is general enough to apply to all cases. Let us now make a list of the memories that are found in living organisms:

(1) The genome is surely a deposit of information and can be regarded as the *genetic memory* of a cell.

(2) The determination state of embryonic development represents a *cell memory*, because it remains in a cell and in its descendants for life, and acts as a deposit of information for all cellular activities.

(3) At the supracellular level we know that there are deposits of information in the nervous system and in the immune system, and it is precisely because of this that we speak of a *nervous system memory* and of an *immune system memory*.

These are the biological memories that we are familiar with, and if it were not for the mathematical model probably we wouldn't feel any need to look for others. According to that model, however, a multicellular system can have a "collective memory", and this does raise the suspicion that a more general memory could exist. More precisely it makes us think about *a supracellular memory to which all the body's apparatuses contribute*, a true *body memory*.

Once the problem is formulated in these terms, it is not difficult to realise that such a memory does indeed exist, because the body plan has precisely the required features. The body plan is a set of characters that defines an animal phylum, and is surely a *supracellular* structure to which all apparatuses contribute. But the body plan is also a *memory*, because it is a structure that appears at an early stage of development, and, from that moment on, it remains in the organism for life, acting as a deposit of information for the three-dimensional pattern of organs and apparatuses. It is as if each organ "remembers" the position and the relative size that it must have in respect to the other organs of the body and behaves accordingly, because the penalty for forgetting them would be swift and exemplary. If some cells abandon their organ and try to establish a colony somewhere else, they are immediately induced by their new neighbours to commit suicide by apoptosis. The memory of the body plan is not an abstract concept but a very basic reality of animal life.

The mathematical model, in conclusion, allows us to add a new biological property to the *Bauplan*, an idea that can be expressed in this way: *a body plan is a supracellular memory*, or *the body plan is the body's memory*. The proof that such an addition is not only new, but also useful, can come of course only from its power to solve real biological problems. One of which is precisely the problem of the Cambrian explosion.

A new model of the Cambrian explosion

The existence of organs and apparatuses in an animal implies the existence of a body plan, and therefore even the most primitive animals (with the possible exception of sponges) had body plans. It is unlikely, however, that the very first animals could already use their body plans as deposits of information, i.e. as supracellular memories. We have seen that the embryonic development of many characters can be realised with two different strategies, a *continuous* mechanism (simpler) and a *discontinuous* one (more complex), and the simpler mechanism is also the one that comes first in the history of life.

In the case of behaviour, for example, a totally instinctive modality is not only simpler but also more *primitive* than a behaviour which is dependent on some forms of learning. More generally, a strategy that is totally determined by genes comes before a strategy that is controlled by genetic and epigenetic factors, and it is likely that this also happened in the evolution of embryonic development.

It is likely, in other words, that the first animals had embryonic developments totally programmed by genes, and that only later did developmental strategies evolve that could also exploit the supracellular information of the body plan. This shift from a continuous (one-phased) development to a discontinuous (two-phased) one would have been a tranformation of enormous importance, and could well correspond to the Cambrian explosion. This is a new hypothesis, and it may be worthwhile to consider its predictions for what happened before, during and after the Cambrian explosion.

(1) *Before the explosion*

The embryonic development of the very first animals was almost totally hard-wired, and this had two important implications. The first is that all embryonic stages were controlled by genes, including the phylotypic stage, and this means that *the body plans were modifiable by genetic changes.* This was the period in which old body plans could be transformed and new body plans could be invented. The second implication is that those animals were *necessarily small* and relatively simple, because there is a limit to the number of characters that can be directly controlled by genes.

(2) *During the explosion*

The shift to an embryonic development that could use, from a certain point onwards, the supracellular information of the body plan allowed *the addition of new developmental stages*, and a longer development could produce a more complex animal, which explains, among other things, the appearance of bigger bodies. It is important to notice that the shift between the two types of development did not require many genetic innovations, because the body plans were *already existing* and provided ample deposits of *potential* information for the spatial pattern of organs and apparatuses. All that was needed was a tranformation of that potential information into actual information, and in principle the employment of already existing instructions can be a very quick process.

(3) *After the explosion*

When the body plan was used as a supracellular memory, embryonic development had a greater number of stages, and therefore produced bigger and more complex animals, but there was a price to pay, because now the body plan could no longer be modified. A set of characters can be a *memory* only if it doesn't change, and this implies that the characters in question must be conserved in all successive generations. (The same is true for behaviour: once a learning-dependent behaviour has been established, learning must be conserved, and successive generations do not return to a behaviour which is totally controlled by genes).

After the Cambrian explosion, therefore, the biological role of body plans changed suddenly and completely: as long as they were not used as information stores, they could be modified at will, but when they became supracellular memories they had to be conserved. New body plans could be produced only by organisms that maintained a continuous type of development, where all stages were totally controlled by genes, and this suggests that such organisms became extinct during or soon after the Cambrian explosion.

We have therefore a new model which can be summarised in this way: *the Cambrian explosion was the transition from a primitive type of development that was totally controlled by genes to a discontinuous type of embryonic development that could also use, from a certain point onwards, the supracellular information of the body plan.*

It may be useful to underline that the model has a number of positive features: (1) it explains why body plans were modifiable before the Cambrian explosion and had to be conserved afterwards; (2) it explains why animals were small before the explosion and could grow to much bigger sizes afterwards; and finally (3) it explains why the explosion took place in a geologically brief period of time.

The model is dependent on the concept of *supracellular memory*, and it may be useful to keep in mind that the general properties of this *collective memory* not only correspond to real biological characters, but can also be simulated by a mathematical model.

The conservation of the phylotypic stage

Von Baer discovered the phylotypic stage of vertebrates at a time when earlier developmental steps were still unknown, and concluded therefore that the very first period of development was devoted to building the body plan. From this he derived the idea that embryonic development proceeds *from the general to the particular.* First it is the characteristics of the phylum that appear, then come the features that specify the class, the order, the family, the genus and the species, and only at this point do the signs appear that distinguish an individual animal from the others.

Within a few decades, however, embryologists discovered that the phylotypic stage is preceded by a number of developmental steps, and that differences between the embryos of the same phylum are clearly visible even *before* the phylotypic stage and not only afterwards (Figure 7.3). The striking similarity which is observed at the phylotypic stage is therefore the result of *two opposite patterns of development*: before the phylotypic stage, the morphological differences between the embryos of a phylum are *decreasing*, whereas after that stage they are *increasing.* The tendency to move "from the general to the particular" was valid after the phylotypic stage, but not before it, and this deprived the idea of the power to explain the whole of development, the very point that von Baer had insisted on.

After the publication of Darwin's *On the Origin of Species*, the entire approach to develoment changed radically, and the idealistic principle proposed by von Baer was replaced by an evolutionary interpretation. The experimental pattern that was expected in this new framework, however, did not change. The developmental stages of embryos are a result of evolution, and since chance variations increase the diversity of organisms, we should observe that differences within a phylum increase in *all* stages of development, and not only after the phylotypic stage.

The pattern which is observed before the phylotypic stage simply cannot be explained by the same evolutionary mechanism that produce the opposite pattern of the other stages, and this is a problem which has never been given a satisfactory answer. Haeckel, for example, claimed that pre-phylotypic stages were simply a result of secondary complications. In the twentieth century, the discovery of genes introduced another major theoretical novelty, and eventually embryonic development came to be seen as the execution of a genetic programme. The stages of development were no longer a *recapitulation* of evolutionary events, but different implementation steps of the genetic programme. Even in this new framework, however, the problem persisted. The evolution of past genetic programmes by chance mutations should have made all stages of development increasingly different, and not only those that come after the phylotypic stage. The opposite pattern of the earlier stages remained a mystery -and could only be attributed to unknown complications.

As we can see, the underlying difficulty is always the same. An evolutionary mechanism based on random mutations is bound to predict divergent patterns in all stages of development, whereas the real patterns are convergent before and divergent only after the phylotypic stage. This has been a crucial problem ever since von Baer's time, but today we have a new model for the Cambrian explosion, and that model does have something new to say about the phylotypic stage problem.

The difficulty encountered by all theories proposed so far is due to the fact that embryonic development has always been regarded as a *continuous* process from fertilized egg to adult, and in this case it is virtually impossible to explain two opposite patterns with the same mechanism. In the framework of the new model, an embryo's development consists instead of *two distinct developments that take place in series:* the first leads to a phylotypic body and exploits only genetic and cellular information; the second leads to an individual body and is also based on the supracellular information of the body plan.

This means that *the end of the first phase of development is also the beginning of the second one*, and in order to achieve this result it is *necessary* that the morphological differences between the embryos of a phylum have two opposite patterns. In the first phase the differences must *decrease* because the embryos of a phylum are building the same body plan and as they approach that target the number of common characters becomes more and more visible. In the second phase, the body plan is no longer an end result, but a starting-point, and from that moment onwards the differences between the embryos of a phylum can increase indefinitely.

The crucial point is the idea that a body plan is *simultaneously* a phenotypic structure and a deposit of information. If information could be transported without three-dimensional structures, there would be no need to conserve three-dimensional patterns, but the information of a body plan is precisely about spatial organisation, and cannot be preserved without the three-dimensional structures which define that organisation. Traditional theories, in conclusion, have regarded the body plan exclusively as a phenotypic structure, not as a deposit of information (a supracellular memory), and it is this which has prevented them from explaining the conservation of the phylotypic stage.

8

SEMANTIC BIOLOGY

Genetics was born in the first years of the twentieth century with the discovery that hereditary characters are carried by molecules that are physically present in chromosomes. These molecules of heredity – *the genes* – are responsible for the visible structures of the organisms but do not enter into those structures, which means that in every cell there are molecules which determine the characteristics of other molecules. In 1909, Wilhelm Johannsen concluded that this distinction is similar to the difference which exists between a project and its physical implementation, and represents therefore a dichotomy of the living world which is as deep as the Cartesian dichotomy between *mind* and *body*.

In order to distinguish the two types of molecules Johannsen called them *genotype* and *phenotype*, but such a dualism was almost universally rejected. At that time it was thought that proteins were responsible for both the visible structures and the hereditary characters, and all biological features seemed reducible to a single type of molecule. The reality of the genotype–phenotype distinction was proved only in the 1940s and 1950s, when molecular biology discovered that genes are chemically different from proteins, and, above all, when it became clear that genes carry *linear information* whereas proteins function through their *three–dimensional structure*.

The genotype–phenotype duality is therefore a dichotomy which divides not only two different biological functions (heredity and metabolism), but also two different physical quantities (information and energy). It is at the same time the simplest and the most general way of defining a living system, and has become the founding paradigm of modern biology, the scheme which has transformed the *energy-based* biology of the nineteenth century into the *information-based* biology of the twentieth. In the end, however, it was the computer that gave a universal legitimacy to Johannsen's duality. The distinction between *software* and *hardware* made people realise immediately the logical difference that divides genotype from phenotype, and gave to Johannsen's dualism a formidable intuitive basis.

All this has undoubtedly been progress, but of course it does not preclude the possibility of an even more general paradigm. And such a possibility became suddenly real in the 1960s, when the discovery of the genetic code proved that in life there are not only *catalysed* assemblies (informatic processes) but also *codified* assemblies (semantic processes). The same logic by which energy and information had been distinguished was suggesting that a new step should be taken in order to distinguish between energy, information and *meaning*. But that new step was not taken. The genetic code was declared a *frozen accident*, an extraordinary feat of nature which took place at the origin of life and was not followed by other organic codes for the remaining 4 billion years of evolution.

In 1981 I argued that the cell is a trinity of genotype, phenotype and ribotype, and in 1985 I proposed the mechanism of *evolution by natural conventions*, i.e. the idea that other organic codes appeared throughout the history of life, but the two proposals did not have any impact. In the 1980s, only Edward Trifonov was also campaigning in favour of other organic codes, suggesting that at least three sequence codes exist in nature in addition to the triplet code, but he too was largely ignored. In the late 1990s, however, things started changing.

In 1996, Redies and Takeichi proposed an *adhesive code* in order to account for the behaviour of cadherines in the developing nervous system, and William Calvin wrote a book entitled *The Cerebral Code*. In 1998, I showed that only organic codes can explain signal transduction and splicing, and Chris Ottolenghi found that various molecular networks are best accounted for in terms of *degenerate codes*. In 1999, Nadir Maraldi extended the idea of signal transduction codes to *nuclear signalling*, and Richard Gordon introduced the idea of a *differentiation code*. And finally in the year 2000 came the fascinating reports of a *sugar code* (Gabius, 2000) and of a *histone code* (Strahl and Allis, 2000; Turner, 2000). The tide is clearly turning, and all indications are that the informatic biology of the twentieth century is going to be replaced by a semantic biology which finally accounts for the existence of organic codes at the very heart of life.

A shift to this new paradigm, however, is bound to take time. Biologists need models to tackle the problems of their research fields, and one cannot ask them to abandon those models simply because they are based on an informatic logic. It is necessary to propose alternative semantic models, and then let the experiments decide. A semantic view of life, a view that takes energy, information and meaning into account, must be able to offer alternative explanations in all fields, and in particular it must be able to propose new models of the cell, of embryonic development and of evolution. Perhaps it is inevitable that the first attempts are vague and imprecise, but there is no alternative, and somehow a first step must be taken. In this chapter, therefore, the arguments of the previous chapters will be used to present the first models of semantic biology, in the hope that this will encourage others to continue the building of the new paradigm.

The semantic theory of the cell

The idea that an embryo is an epigenetic system goes back to Aristotle and, after the interlude of preformationism, has become an integral part of modern biology. The idea that the cell too is a system that increases its own complexity is, on the contrary, completely new. Epigenesis has never been *explicitly* named among the fundamental properties that define the cell (see Appendix), even if the experimental data that support this conclusion have been known for a long time: the linear information of the genotype does not contain a complete description of the phenotype, even at the cellular level, which means that every cell is a system where *the phenotype is more complex than the genotype.*

Epigenesis exists therefore even at the most fundamental level of life, but acknowledging this reality unfortunately is not enough. We need to understand how a system manages to become more complex, otherwise the word *epigenesis* becomes a mere label that is conveniently used only to cover up our ignorance, just like *vital force* in the past. We need, in other words, a *mechanical* model of epigenesis in order to understand it. Luckily today we do have such a model, and we can at least try to apply it to the cell. The model shows that a reconstruction from incomplete information is possible only if the system in question contains one or more memories, and in the case of the cell this amounts to a first general conclusion: *all cells must have one or more organic memories.*

The model has also shown that new information about the system can indeed appear in the memory space, but this information can be transferred to the real space only by codes of correspondence between the two spaces. And this gives us a second general conclusion: *all cells must have one or more organic codes.* The result is that every cell must have (1) organic structures, (2) organic memories and (3) organic codes. The organic structures make up the *phenotype*; the organic memories store at least the linear information of the *genotype* (but can also be the seat of contextual or three-dimensional information); and the organic codes include at least the genetic code of the *ribotype* (but can also be other organic codes).

The mathematical model, in short, suggests that a link exists between the three new hypotheses on the cell that have been illustrated in the previous chapters: (1) the idea that a cell is an epigenetic system because its phenotype is more complex than its genotype (Chapter 1); (2) the idea that all cells have organic memories and organic codes (Chapter 4); and (3) the idea that a cell is a system made up of genotype, ribotype and phenotype (Chapter 5). The important point is that these three hypotheses are mutually compatible, and can therefore be combined into a single proposition that will be called *the semantic theory of the cell*:

"The cell is an epigenetic system made of three fundamental categories (genotype, ribotype and phenotype) which contains at least one organic memory (the genome) and at least one organic code (the genetic code)."

It must be underlined that the *minimal* requirements of this definition (a single memory and a single organic code) can be attributed only to very primitive cells, because even bacteria are more complex systems. Prokaryotic cells have at least one other code (for signal transduction), while unicellular eukaryotes exploit at least two other codes (signal transduction codes and splicing codes).

A second important point is that the semantic theory of the cell is not merely an addition of the ribotype concept to the classical duality of Johannsen. The self-replicating machine described by von Neumann is also a system which introduces a *ribotype* (a universal constructor) between genotype and phenotype (Mange and Sipper, 1998), but is not a valid model of a living cell, because von Neumann's genotype must contain a *complete* description of its phenotype. Von Neumann's machine, in other words, is not an epigenetic system. The semantic theory, in contrast, has its very foundation in the idea that a cell is an epigenetic system, and states that organic codes and organic memories are indispensable precisely because only they can make a phenotype more complex than its genotype. If the living cell is an epigenetic system, in conclusion, then it is bound to have organic codes, and therefore it is bound to be a semantic system.

The semantic theory of embryonic development

After fertilisation and a first round of cell divisions (cleavage), the early embryo begins what is probably the most important phase of its development (gastrulation) by separating the cells that remain in contact with the outside world (ectoderm) from those that bury themselves inside the body (endoderm and mesoderm). The ectoderm is the first "skin" of the embryo, and the primary purpose of any skin is to form an impermeable barrier around the body (if water could enter and exit freely, the size and the shape of a body would change erratically according to the surrounding degree of humidity).

The ectoderm ensures that the inner space of a body is sharply distinct from the outside world, and all forms and shapes that we find in that inner space are entirely due to *endogenous* processes of three-dimensional organisation. And the same is true for every single cell. The plasma membrane controls virtually everything that is passing through, and is for a cell what ectoderm is for an embryo: the structure that sharply divides the inner space from the outer world, thus giving the inner system the freedom to build its own structures in complete autonomy.

In every organism we find therefore *two biological spaces* and *two kinds of three-dimensional organisation:* the inner space of the body where cells are spatially organised in tissues and organs, and the inner space of every single cell where organic molecules are spatially organised in organelles and subcellular structures. It is important to notice that the two spaces and the two types of spatial organisation are totally different. Even if the organelles of a cell are in some way comparable to the organs of a body, the mechanisms that build bodily organs are utterly different from those that assemble subcellular organelles (at the organs' level, for example, there is nothing comparable to the mechanism of dynamic instability of the cytoskeleton).

The one and only parallel that can be made between a cell and the multicellular body that contains it is that both have the same genotype, and that in both cases the phenotype is more complex than the genotype. They both are, in other words, epigenetic systems that must solve – with different mechanisms – the same problem of reconstructing their structures from incomplete information.

In the course of embryonic development, therefore, two different reconstruction processes are taking place *in parallel*, one at the cellular level and the other at the organism level. These two parallel processes, furthermore, are both *discontinuous*, because in each of them there is a *crucial event* which marks the end of one phase and the beginning of a new one. For a developing cell the crucial event is *determination*, the trauma that forever fixes its destiny; for an embryo it is the appearance of the *body plan*, the pattern that determines for life the three-dimensional organisation of the body's organs.

Determination and body plan are both arrangements of structures in space which are conserved in time, and therefore behave as "organic memories". They are veritable stores of information, but in this case it is three-dimensional information that they are carrying. The problem is that the concept of three-dimensional – or *contextual* – information is difficult to deal with, and often it is confused with the idea of three-dimensional structure *tout court*. This is why it is important to have mathematical models that help us to distinguish clearly between space-structures and space-information.

We have already seen that the reconstruction model makes a sharp distinction between real-space matrices and memory-space matrices, and suggests that embryos too have equivalent types of three-dimensional structures: the "real" structures of the phenotypic body, and the "memory" structures of the body plan. According to the reconstruction model, therefore, during embryonic development there is not only a development of phenotypic structures, but also a parallel development of memory structures. It should be noticed, however, that these two processes are sharply asymmetrical.

The phenotypic development of a body is practically a continuous process, while the development of contextual information is divided into two very different phases by the discontinuity of the body plan. In the first period, when the body plan does not exist, there is contextual information only at the cell level, and memory structures are confined to the cell. In the second phase, in contrast, the body plan becomes a source of contextual information above the cell level, and the embryo's development takes place in a totally new reference system. From this moment on, the body plan becomes a new intermediary between genotype and phenotype, and since it is shared by all members of a phylum, we can call it the *phylotype*, or the *phylotypic body*. The reconstruction model, in short, allows us to conclude that a multicellular organism consists of three fundamental categories – *genotype, phylotype and phenotype* – and that embryonic development consists of two different developments which come one after the other, i.e. *in series*. The first is the development of the *phylotypic* body; the second is the development of the *individual* body.

At this point we can combine the above concepts in a single model, and obtain *the semantic theory of embryonic development*:
"Embryonic development is a sequence of two distinct processes of reconstruction from incomplete information, each of which increases the complexity of the system in a convergent way. The first process builds the phylotypic body and is controlled by cells. The second leads to the individual body and is controlled not only at the cellular level, but also at the supracellular level of the body plan."

The mind problem

The study of language became a science when its ideas started to be submitted to experimental tests, and one of the first achievements of the new science has been described with great clarity by Massimo Piattelli-Palmarini:

"Just as there is a naive physics whose intuitions are subverted by true physics, so there is a naive theory of language which is easily dismissed by experimental data. This theory states that a child learns new words by listening to the sounds which accompany the actions performed by adults or the objects which are presented to him. In reality, as Paul Bloom has noticed, no mother, on coming home, tells her child: 'Now I am opening the door. Now I am hanging up my coat. And now I am coming towards you to give you a kiss'. Rather, in doing all these things, she is likely to say something like 'How was your afternoon? Did you play with Maria? Did you brush your teeth?' And normally a statement is going to be uttered just when the corresponding action is not taking place. 'Sleep' will be said when the child is not sleeping."

Children must clearly be born with mental rules which allow them to *interpret* what adults are saying, and luckily today we know of many examples which prove their existence. Susan Carey and Nancy Soja, for example, were able to illustrate some of them by describing how two-and-a-half-year-old children handle imaginary names. A T-shaped metal object (a hydraulic joint) was given the name "blinket", while a piece of dough was called "dax", and then the children were invited to find other "blinket" and other "dax" objects in the room. It turned out that they identified as "blinket" any T-shaped object, even if made of wood, cardboard or plastic. The name "dax", instead, was given only to dough-made objects, irrespective of their shape, and was never used for pieces of wax or jelly, even when these had the same shape and size of the original piece.

Clearly the children deduced that "blinket" meant a particular shape and not a material substance, while "dax" referred to a material substance and not to a shape. These are abstract hypotheses, and children must have inborn mental rules in order to perform such complex operations.

Without a theory of the mind it is impossible to explain how a language is learned, and we must conclude that a child has an *inborn mind*, a set of mental rules and mental objects which allow him to interact with the external world. It is known, furthermore, that a child can learn any language whatsoever, and this means that the inborn mind must contain a set of rules which apply to all languages, a set that Noam Chomsky (1965, 1972) has called *universal grammar*.

Against the theory of inborn ideas there has been proposed, for centuries, the opposite view that the mind of a newborn child is a *tabula rasa* where only experience, like a writing hand, can begin to leave marks, and the whole debate about human nature has traditionally been centred on the opposition between hereditary characters and environmental factors.

This classic contrast between *nature* and *nurture*, between heredity and environment, between genotype and phenotype, has also dominated the theories of language, and the recent discovery of inborn mental rules appears to have suggested a sort of compromise solution: *up to the moment of birth mental development is under genetic control, while after birth it becomes dependent upon environmental stimuli.* With the terminology that has been adopted today, the *universal grammar* would be determined by the genes, whereas the *individual grammar* would be almost entirely a product of the environment.

In reality, such a solution is not at all satisfactory, either from a philosophical or from a biological point of view. What does one actually mean with the statement that genes control the rules of the universal grammar but not those of the individual one? Perhaps that genes contain *all* the instructions that make up the universal grammar? That the environment does actually deliver all the instructions which shape the individual grammar?

The real problem, here, is that the development of the mind is characterised by a *convergent increase of complexity* both before and after birth, and the present compromise solution on language doesn't help us in the least to understand this fundamental process. The stumbling-block is that any dualistic scheme such as heredity–environment, or genotype–phenotype, is a priori incapable of solving the problem because any machine, or system, that works on the basis of a software–hardware logic cannot increase its own complexity. As in the cases of the cell and of embryonic development, therefore, we need a new theoretical framework for mental development, even if this means that we have to abandon the *nature–nurture* scheme that has been imposed for centuries on our approach to the human mind.

The semantic theory of mental development

It has been held for centuries that mind and body are divided by an unbridgeable gulf, but in reality there is no actual proof that they develop with totally different mechanisms. There are, on the contrary, some intriguing common features in their developments. We have seen that a universal grammar must appear in a very early phase of mind development, and in that phase we can rightly say that a child has a *species-specific mind*, or a *specietypic mind*, because that mental state is shared by all members of our species.

This suggests immediately that the *specietypic stage* of mental development is comparable with the *phylotypic stage* of embryonic development. In both cases, it is necessary that all members of a taxonomic group pass through a common phase of development before they begin developing individual characteristics. Even mental development, in other words, is a sequence of two processes: one that builds the specietypic mind, and the other that goes on from that stage and builds the individual mind.

We have seen, furthermore, that the phylotypic stage of embryonic development is very short, but the body plan that is built in that brief interval remains for life, and acts as an organising centre for the individual body. And the same is true for the mind: the phase of the specietypic mind is transient, but the universal grammar that is built in that brief time does not disappear, and becomes the organising centre of the individual mind. We have also seen that, in the language learning period, a child actually encounters only an extremely limited and erratic sample of words and phrases, and yet, in the end, all children in a population learn the same language, and spontaneously invent countless rules that nobody taught them. There is an enormous gap between environmental stimuli and final result, between input and output (the so-called *poverty of the stimulus*), and we can conclude therefore that *even mental development is a reconstruction from incomplete information.*

Let us now summarise the main points: (1) there is a phylotypic stage in the development of the body and a specietypic stage in the development of the mind; (2) there is a body plan in the development of the body and a universal grammar in the development of the mind; (3) there is incomplete information in the development of the body and incomplete stimuli in the development of the mind. If we put these conclusions together, we obtain a semantic model of mind development just as we did for embryonic development:

"Mental development is a sequence of two distinct processes of reconstruction from incomplete information each of which increases the complexity of the system in a convergent way. The first process builds the specietypic mind (the universal grammar), while the second leads to the individual mind."

It is important to notice that these conclusions are not obtained by mere *analogies* between mind and body, because in both cases the starting-point has been a well-known set of experimental data. In mental development, the basic fact is the *poverty of the stimulus*, a reality which is widely documented but which has never been explained. The reconstruction model is the first rational explanation of that phenomenon, and this does qualify it as a true scientific hypothesis. There is also another consequence of the model that should be outlined. Richard Dawkins (1976) had the great merit of introducing into biology the concept of *mental genes* or *memes*, but the semantic theory requires something more than that, because multilevel mental reconstructions cannot be performed from memes alone. Between memes and mind, in other words, there must be intermediate mental structures, such as *mental cells* and *mental organs*, in order that complex reconstructions can actually be implemented. As we can see, the semantic theory not only offers new explanations for unsolved experimental problems, but also predicts new mental structures that one day could be discovered.

Artifacts and natural selection

Before the appearance of human language, the Earth was inhabited
by two kinds of objects: the creatures of the living world, and the
inanimate objects of the physical world such as stones, clouds, rivers,
lightning and volcanoes. With language, a third kind of object came
into being, and our planet started to be populated by human artifacts
such as knives, wheels, clocks, books and windmills.

Artifacts are inanimate objects, and one could expect that their
diffusion in nature is described by equations such as those that apply
to stones and clouds, but things turned out to be very different. From
a mathematical point of view, it has been discovered that artifacts
behave exactly as living organisms.

From the 1970s onwards, Cesare Marchetti and other system
analysts have studied thousands of artifacts, and have discovered that
their behaviour is described by the same equations that Lotka and
Volterra found for the behaviour of predators and prey. The growth
pattern of cars, for example, is a logistic curve. Cars spread in a market
exactly as bacteria in a broth or rabbits in a prairie. Cultural novelties
diffuse into a society as mutant genes in a population, and markets
behave as their ecological niches. But why?

The answer is that artifacts originate in the human mind as *mental
objects*, and afterwards are turned by man into *physical objects* (this is
true even for a poem which must become ink on paper or sound-
waves in air). Artifacts have therefore a *genotype* and a *phenotype*.
The genotype of the pen that I am writing with is the idea that was
born in its inventor's mind, and which was replicated countless times
in the blueprints of its mass production. The real pen that my hand
is holding is an inanimate object because it is a pure phenotype, a
phenotype which is physically separated from its genotype, but I
could never understand its existence if I didn't keep in mind that
this physical object came from a mental object.

Human artifacts, in short, have a genotype and a phenotype, and
this qualifies them as "organisms", but they are organisms of a very
peculiar breed, because their phenotypes are totally dissociated from
their genotypes. They truly are a new form of life, and their appearance

on Earth was a real episode of macroevolution.

The physical separation between genotype and phenotype has an extraordinary consequence, because mental genotypes can be directly instructed by mental phenotypes, and this means that cultural heredity is based on a transmission of acquired characters. Cultural inheritance, in other words, is transmitted with a Lamarckian mechanism, whereas biological inheritance relies on a Mendelian mechanism which is enormously slower. As a result, cultural evolution is much faster than biological evolution, and almost all differences between biology and culture can be traced back to the divide that exists in their hereditary mechanisms. The discovery that human artifacts (i.e. cultural phenotypes) obey the Lotka–Volterra equations has two outstanding consequences. The first is that selection accounts for all types of adaptive evolution: *natural selection is the mechanism by which all phenotypes – biological as well as cultural – diffuse in the world.*

The second consequence is yet another testimony that an enormous divide exists between phenotypic evolution and molecular (genotypic) evolution, because it means that selection is independent from heredity: *natural selection does not depend on a particular hereditary mechanism because it works both with the Mendelian mechanism of biological organisms and with the Lamarckian mechanism of cultural objects.* Natural selection explains well how human artifacts diffuse into a market, but this has nothing to do with the origin of ideas in the human mind. And exactly in the same way, natural selection explains well the diffusion of living creatures in their ecological niches, but has no power on the origin of biological novelties in the genotypes' universe. Selection acts on what already exists, and knows nothing of the creation of life.

The semantic theory of evolution

According to modern biology there are only two codes in nature: the genetic code – which appeared on Earth with the origin of life – and the human codes of cultural evolution, which arrived almost 4

billion years later (Figure 8.1). This implies that no other code came
into existence for nearly 4 billion years, and therefore that biological
evolution never produced any other organic code after the origin of
life. According to modern biology, in other words, from the first
cells onwards, biological evolution took place only with informatic
mechanisms, and not with semantic ones.

The discovery of other organic codes is bound therefore to make
us think again about the mechanisms of evolution. When biologists
will finally realise that phenomena such as splicing and signal
transduction are based on true organic codes, exactly as translation is
based on the genetic code, they will also realise that the history of life
cannot be understood without the history of organic codes.

Before studying this history, however, we need to address a
methodological problem. The evolution of any organic code is a
historical process, and either the beginning or the end of that
process could be taken as the "origin" of the code. Here, however,
it will be assumed that the origin of a code corresponds to the
appearance of a complete set of rules, i.e. to the end of its *primary*
evolution. This choice does have some drawbacks, especially in the
case of splicing. If we say that splicing codes appeared with the
first eukaryotes, some 2 billion years after the origin of life, it could
rightly be objected that some splicing phenomena were probably
much older than that, and could have been present at the very
beginning, which is certainly possible. We need therefore to justify
the above choice, and the justification is this: *the origin of an organic
code is the appearance of a complete set of rules, because when that
happens something totally new appears in nature, something that
did not exist before.*

In the case of the genetic code, for example, we have already seen
that its rules could have appeared one at a time in precellular systems,
because each of them could give a contribution to the development
of those systems. When a *complete* set of rules appeared, however, it
was something totally new on Earth: what came into being was
biological specificity, the most fundamental of life's properties. That
event marked the origin of exact replication, the birth of the first true
cells, and it is proper therefore to say that the origin of life coincided

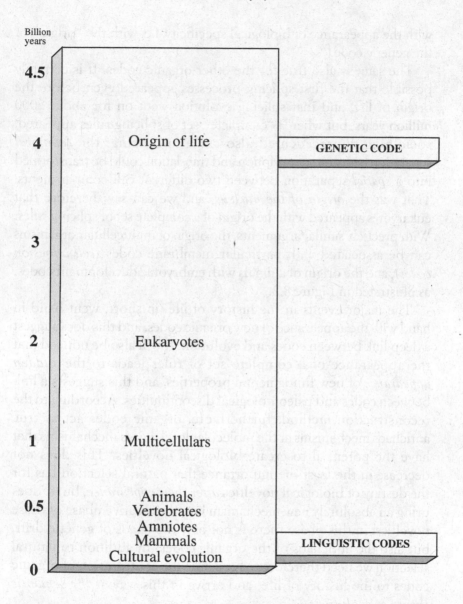

Figure 8.1   According to modern biology textbooks, the genetic code of the cell and the linguistic codes of culture are the only codes that exist in nature, which implies that almost 4 billion years of biological evolution did not produce any other organic code.

with the appearance of biological specificity, i.e. with the "origin" of the genetic code.

The same is also true for the other organic codes. It is certainly possible that the first splicing processes appeared even before the origin of life, and that splicing evolution went on for about 2000 million years, but when a "complete" set of splicing rules appeared, something unprecedented also came into being: the *temporal* separation between transcription and translation could be transformed into a *spatial* separation between two different cell compartments. That was the *origin of the nucleus*, and we can say therefore that eukaryotes appeared with the origin of a complete set of splicing rules. With precisely similar arguments, the origin of multicellular organisms can be associated with particular membrane codes (*cell adhesion codes*), and the origin of animals with embryonic development codes, as illustrated in Figure 8.2.

The major events in the history of life, in short, went hand in hand with the appearance of new organic codes, and this does suggest a deep link between codes and evolution. It will also be noticed that the appearance of a complete set of rules leads to the *sudden appearance* of new fundamental properties, and this suggests a link between codes and paleontological discontinuities. According to the reconstruction method, furthermore, organic codes act as true antichaos mechanisms at the molecular level, i.e. as mechanisms that have the potential to create biological novelties. This does not decrease in the least the importance that natural selection has for the destiny of biological novelties *after their appearance*, but it does bring an absolutely new mechanism into the creative phase of those novelties: in this phase there is not only the chaos of genetic drift, but also the antichaos of the organic codes. In addition to natural selection we need therefore to recognise the contribution of organic codes to the history of life, and arrive in this way at *the semantic theory of evolution*:

"The origin and the evolution of life took place by natural selection and by natural conventions. The great events of macroevolution have always been associated with the appearance of new organic codes."

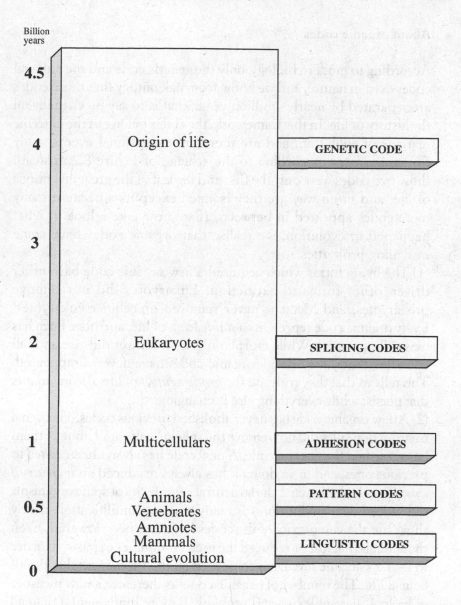

Figure 8.2   According to semantic biology, many organic codes exist in life and their appearance on Earth marked the great historical events of macroevolution. The actual number of organic codes is likely to be much higher than the few examples reported above.

About organic codes

According to modern biology, only the genetic code and the cultural codes exist in nature, and we know from paleontogy that those codes are separated by nearly 4 billion years, that is to say by virtually all the history of life. In this framework, the codes belong to the extreme margins of evolution, and are therefore exceptional events, truly *frozen accidents*. According to the scheme of Figure 8.2, instead, those two codes were only the first and the last of the great inventions of life, and in no way are they isolated exceptions because many more codes appeared in between. If we now take a look at what happened in evolution, we realise that organic codes have some intriguing properties. ·

(1) The living forms which acquired a new organic code have never driven other forms to extinction. Eukaryotes did not remove prokaryotes, and Metazoa never removed unicellular eukaryotes. Every organic code represents a *stable* form of life, and once born has never disappeared. While morphological structures did rise and fall countless times, the "deep" organic codes have never disappeared. This tells us that they truly are the *fundamentals* of life, the invariants that persist while everything else is changing.

(2) A new organic code has never abolished previous codes. The signal transduction rules did not remove the splicing rules, and none of them has abolished the genetic code. A new code has always been *added* to previous ones, and in so doing it has always produced an *increase of complexity* in the system. The structural complexity of some organisms did indeed decrease in time, as many cases of simplification clearly show, but the complexity of the codes has never been lowered. Even the animals that lost or reduced the greatest number of parts, in order to lead a parasitic life, have conserved all the fundamental codes of animal life. The number of organic codes is therefore a new measure of biological complexity, and probably is more fundamental than all other parameters which have been proposed so far.

(3) The genetic code is present in all living creatures, but the other organic codes appeared in increasingly smaller groups, thus giving rise to a veritable "pyramid" of life. The greater the number of codes,

the smaller the number of species that possess them, as shown in Figure 8.3. Such a pyramid could give the impression that evolution is somehow "oriented", but in reality there is no need to explain a perfectly natural outcome with the intervention of additional and unnecessary guiding principles.

(4) Even if the evolution of an organic code could take an extremely long time, the "origin" of a complete code is a sudden event, and this

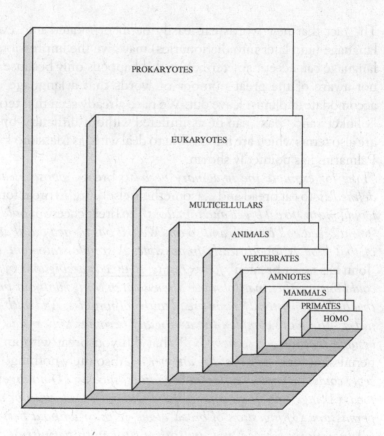

Figure 8.3   The evolutionary novelties which characterised more complex organisms appeared in a temporal sequence (vertical axis) in progressively smaller subgroups of pre-existing forms (horizontal axis), thus creating the impression of an oriented process.

means that the great evolutionary novelties associated with that code appeared suddenly in the history of life. This is a new explanation of the discontinuities that paleontology has documented, but it is not an *ad hoc* explanation, because the observed effect is a direct consequence of the codes' properties.

The language model

The fact that new words can easily be incorporated into everyday language (and later into dictionaries) may give the impression that a language can accept any term, but this happens only because we are not aware of the great number of words that a language cannot accomodate and must leave out. We have already seen that terms like "blinket" and "dax" can be assimilated without difficulty, but there are also terms which are impossible to deal with, as Massimo Piattelli-Palmarini has pointedly shown.

"Take for example the imaginary verb 'to breat', *whose meaning is defined as* 'to eat bread and ... (something else)'. *So,* 'to breat tomatoes' *would mean* 'to eat bread and tomatoes', 'to breat cheese' *would signify* 'to eat bread and cheese', *and so on. Well, it has turned out that a verb of that kind is linguistically impossible. A simple statement such as* 'John has never breaten', *for example, is already ambiguous because it could mean either that John has always eaten things without bread, or that he has consistently eaten bread with nothing else. And with phrases just a little more complex the ambiguity becomes unbearable. Let us assume that we ask somebody* 'What did you breat without asking permission for?" *and that the answer is* 'Absolutely nothing!' *Such a reply could have many different meanings, such as: (1) no ingestion of food; (2) ingestion of all sorts of food without bread, with or without permission; (3) ingestion of bread alone, with or without permission; (4) ingestion of bread and any other stuff with permission; (5) no ingestion of bread and other food items with permission for bread but not for the other items; (6) no ingestion of bread and other food items with permission for the other items but not for bread. A verb such as this, despite its abstract banality, is mentally unworkable. It is*

unlearnable and unusable and cannot therefore enter into any natural language. No child, in any linguistic community, would ever attempt to give such a meaning to a new verb that he happens to encounter for the first time in a conversation."

This and many similar examples prove that the inborn mind of a child submits any new term to an unconscious "acceptability test", and what is actually scrutinised first is not the content of the new term (the information that it is delivering) but its ability to play *the rules of the game.* What matters, before anything else, are not the individual characteristics of the new term but its *group properties.*

Could this lesson have something to say in biology? Up until now, biologists have only looked for individual features in genes and proteins, not for group properties, which is understandable because the latter are much more difficult and elusive. The main obstacle is that we do not have a mathematical expression for meaning, and this implies that semantic processes cannot be investigated with quantitative methods. It is possible, however, that the group properties that linguistics is discovering may, one day, turn into a good model, or at least provide practical guidelines for biological group properties. The study of language is rapidly growing into a full science, and perhaps in the future we may be able to prove formally, among other things, why it is that a verb like "breat" is linguistically impossible.

Even today, however, linguistics can help. Just as the artificial selection practised by farmers and breeders was a powerful source of inspiration for Darwin, so today language can give us an illuminating model for the history of life. The most important lesson is that language evolution was a combination of two parallel but different processes – evolution of words and evolution of grammatical rules – and this is a fitting model for the two different mechanisms of biological evolution that are proposed by the semantic theory.

The Individuality Thesis

Language is full of terms that we use both in a concrete and in an abstract sense. With the term *chair,* for example, we can refer to the concrete chair that one is sitting on, as well as to the "idea" of chair, the abstract concept that allows us to recognise countless different objects as chairs, even if nobody can actually sit on the idea of chair.

We also have words that can identify groups of objects, for example armies and species, and in these cases we normally employ different terms for different types of groups and subgroups. Armies, for example, are made of platoons, companies, regiments and divisions. Animals are classified into taxa such as species, genera, families, and so on. We distinguish, in other words, between "individuals" and "groups of individuals", or "classes", where classes are *abstract* entities while individuals are *concrete* bodies.

Language has greatly benefited from the ability to name both concrete and abstract things, but this privilege has been a sore source of trouble for scholars from time immemorial. Individuals and classes are things that surely exist in the world of language, *but do they also exist in the real world?* Here people have split into two schools. According to the "nominalists", only individuals are real, while classes are mere names that we give to collections of individuals. According to the "realists", instead, abstract things must have an existence of their own, otherwise we could not use them to grasp the world. We could not have mathematics, for example, if numbers didn't have some kind of "reality" and were only a figment of our imagination. This longstanding debate has not disappeared with the advent of modern science. If anything, it has re-emerged with greater force, especially in biology, where it has become known as "the species problem".

A species is a collection of individuals, i.e. a class, but it is also a concept that seems to have been devised to drive us mad. If we accept the nominalist position (that only individuals are real), we must conclude that species, being classes, are abstract groups, not real things, and if they are not real they cannot evolve. This is enough to make us embrace the realist position, and say that species are real, but this turns into another headache, because now we have to explain how on earth can an abstract thing like a class be abstract and real at the same time. One can almost hear the nominalists' grin: *"It's either abstract or real, you cannot have it both ways."*

At this point we keep calm, take a deep breath, and sweep

everything under the carpet: this is how good old common sense has saved biologists from going insane. Ever since Darwin, the species problem has been safely tucked away, and we have just learned to live with that skeleton in the cupboard. Until 1974, that is, when Michael Ghiselin opened the cupboard and told everybody that the nightmare was over. The solution is that the nominalists are right in saying that only individuals are real, but everybody was wrong in saying that species are classes. Species are real, and therefore they are not classes: they are individuals.

This is Ghiselin's *Individuality Thesis* (also known as the *Radical Solution*), and one could be forgiven for thinking that there must be a trick, somewhere. Except that there is no trick, because Ghiselin has not just said that species are individuals. He has proved it.

We can read the full-length proof in *Metaphysics and the Origin of Species* (1997), where it starts from Aristotle's ten ontological categories. Ghiselin is not very happy with this time-honoured scheme, because it looks more a list than a classification, and so he goes on and (via the Stoics and Porphyry) comes to Kant's twelve categories of pure understanding, only to find out that these are epistemological, not ontological, categories. So Ghiselin goes back to Aristotle and transforms his list into a proper classification by giving a primary role to four categories and a secondary role to the other six (with a little reshuffling of the terms).

The result is that now we have a reference system where we can finally face the problem of defining the term *individual* in a rigorous way. At which point Ghiselin takes us through a step-by-step anatomical dissection of that apparently simple word until we come up, in the end, with a list of six criteria of individuality: (1) non-instantiability, (2) spatio-temporal restriction, (3) concreteness, (4) not functioning in laws, (5) lack of defining properties, and (6) ontological autonomy.

This is the basis of Ghiselin's Individuality Thesis: species are supraorganismal individuals because they conform to all six criteria of individuality. The conclusion is that we become fully aware that *"Biological species are, ontologically speaking, individuals in the same fundamental sense that organisms are, and that organisms stand to species as parts to wholes, not as members to classes."*

Later on, Ghiselin became increasingly interested in extending the Individuality Thesis from species to cultures. He was puzzled by the fact that *"There are some very strong analogies between biological species and certain cultural units, especially languages. But there are also important differences, and identifying wholes and their parts can be difficult."* He expressed these feelings in a paper entitled "Cultures as supraorganismal wholes" (2000), where he is clearly tempted by thoughts like *"Culture would be like Life"*, *"cultures would be like biological species"*.

These are ideas that go back at least to 1869, when Schleicher expressed a view which, according to Ghiselin, amounts to saying that *"Languages evolve very much as species do, and similarities between the two are most striking."* And yet, today, most biologists declare that the similarities are only superficial, that the differences are too many and too deep: there is no sex in cultures, no Mendelian heredity, no tissues and organs, no clear embryonic development. Everything looks different, deep down, we are told.

But they haven't looked deep enough, and above all they have looked for biological features in cultures, not for cultural (conventional) features in biology. Indeed, if life evolved not only by natural selection but also by natural conventions, then at the deepest level of all, at the very heart of fundamental change, there would be codes in life as there are codes in culture. Which would mean, among other things, that cultures are indeed "supraorganismal wholes" as species are.

The development of semantic biology

Today, virtually all biology books speak of the genetic code, but none mentions signal transduction codes or splicing codes. Why? Perhaps a brief historical summary may help us to understand. In the 1950s it became clear that protein synthesis requires a transfer of information from nucleic acids to proteins, and people realised that such a process must necessarily use a code. The existence of the genetic code, in other words, was predicted *before* doing the experiments that actually discovered it, and the results of those experiments were correctly interpreted as a proof of the code's existence.

In the case of signal transduction, in contrast, the experiments were planned from the very beginning as a means of studying the biochemical steps of the phenomenon, not as a search for codes, and the biochemical reactions of that field were regarded a priori as normal catalysed processes not as codified processes. *No code had been predicted, and so no code was discovered.* Even if the experimental results of signal transduction can be understood only by admitting the existence of organic codes, it is a historical fact that no one looked for transduction codes, and that is why no book mentions them.

As for splicing codes, the very existence of splicing came as a totally unexpected surprise, and it seemed natural to concentrate on the biochemical steps of the phenomenon before attempting any theoretical interpretation. This meant that splicing was labelled from the very beginning as a catalysed process, and has been studied as such ever since.

As we can see, organic codes can be discovered only if we are looking for them, and we can look for them only if we believe that they *can* exist. In order to build a semantic biology, therefore, the first step is a new mental attitude towards nature, even if this will probably be possible only with a new generation of molecular biologists.

When the theoretical climate is favourable, however, priority will necessarily go to experiments. It is important, in fact, to avoid the opposite extreme of believing that everything can be explained by invoking *ad hoc* codes, or by saying that all biochemical reactions are, more or less indirectly, codified processes. The second step towards semantic biology will be therefore a stage of experiments and discovery.

When the accumulation of data makes semantic phenomena familiar, however, a new phase will necessarily begin, since biologists will have to face the problem of accounting for the very existence of organic codes in nature. It is unlikely that an answer may come from a biology which has systematically ignored these codes, and which accepted the genetic code only because it could not do otherwise.

Perhaps biologists will have to consult artificial-life engineers, and will be told that the existence of organic codes goes hand in hand with the existence of organic memories. And this will give them other problems to think about: Do organic memories exist? Where are they? How many are there? In order to answer these questions, there will have to be a third phase of development, and attention will return once more to experiments.

Eventually, however, even organic memories will become familiar, and biologists will start asking themselves why life has organic codes and organic memories. Why do they exist? Are they really necessary? And necessary for what? This will be the last step in the building of semantic biology. In the end, biologists will discover that organic codes and organic memories are the sole instruments which allow a system to increase its own complexity, and will understand that this is the most fundamental property of all living creatures, the very essence of life.

9

A BRIEF SUMMARY

Semantic biology has been developed in stages since the 1970s. The mathematical papers appeared in 1974 and 1987 (Barbieri, 1974a, 1974b, 1987). The first biological paper was "The ribotype theory on the origin of life" (Barbieri, 1981) and the first general theory (the concept of *evolution by natural conventions*) was proposed in *The Semantic Theory of Evolution* (Barbieri, 1985). The idea that splicing and signal transduction are based on organic codes was introduced much later (Barbieri, 1998), and so was the term *semantic biology* (Barbieri, 2001). There are at least five new concepts in this biology (*ribotype, organic codes, organic memories, reconstruction from incomplete information* and *evolution by natural conventions*), and they have all gone unnoticed for a long time. Something similar happened to the ideas of Edward Trifonov, who has been calling attention to *sequence codes* since 1988, and perhaps that was not a coincidence. Things however seem to be changing. The discoveries of the *sugar code* (Gabius, 2000) and of the *histone code* (Strahl and Allis, 2000; Turner, 2000) have made some impact, and there is a growing awareness that real organic codes do exist in nature. A parallel development has also taken place in philosophy and in linguistics. In 1963, Thomas Sebeok proposed that semiotics must have a biological basis, and has campaigned ever since for a more general approach which today is known as *biosemiotics* (Sebeok, 2001). In the long run, the place of organic codes in nature is bound to be acknowledged, and biology will need a proper theoretical framework for them. Today we only have a preliminary outline of that framework, and this last chapter is going to underline it by showing that semantic biology can be summarised in eight propositions. More precisely, it can be expressed by *four general principles* and *four biological models*.

The first principle

Embryonic development was defined by Aristotle as an *epigenesis*, i.e. a chain of one genesis after another, a step by step generation of new structures, and, apart from the brief interval of *preformationism*, this view has been endorsed throughout the history of biology, and still holds good. Despite its popularity, however, epigenesis is not an easy concept to handle, and for most practical purposes it is convenient to define it as *the property of a system to increase its own complexity*. More precisely, epigenesis can be defined as a *convergent* increase of complexity, in order to emphasise that the oriented character of embryonic development is qualitatively different from the *divergent* increase of complexity that may take place, for example, in evolution.

The historical association of epigenesis with embryonic development has been so close that the two terms are sometimes taken as synonymous, and this has been unfortunate because it has probably prevented biologists from realising that a convergent increase of complexity is a universal feature of life. The definitions of life which have appeared in the last 200 years (starting with Lamarck's entry) have produced a long list of supposedly essential characteristics (heredity, metabolism, reproduction, homeostasis, adaptation, autopoiesis, etc.), but none of them has explicitly mentioned epigenesis (for a list of such definitions see Appendix).

The first principle of semantic biology is precisely this: *epigenesis is a defining characteristic of life*. Any living organism is a system that is capable of increasing its own complexity. Even single cells, for example, can be defined as systems where the phenotype is more complex than the genotype (Barbieri, 2001).

Modern biology has already acknowledged that complexity is at the very heart of life, but semantic biology goes further than that. It states that what is crucial to life is not complexity as such, but the ability to produce a *convergent increase* of complexity. The first principle of semantic biology, in short, is nothing less than a new definition of life.

The second principle

Complexity has a straightforward intuitive meaning (it is the opposite of simplicity), but its scientific history is littered with the corpses of discarded definitions. There simply is no hope of achieving a general consensus on a comprehensive definition of complexity, and this implies that any attempt to give a mathematical formulation to the problem of epigenesis is apparently crippled at the very beginning by lack of a definition.

The second principle of semantic biology has the purpose of cutting the Gordian knot of complexity by formulating the problem of epigenesis without any explicit reference to it. More precisely, the principle states that achieving a convergent increase of complexity is equivalent, to all practical purposes, to *reconstructing a structure from incomplete information*.

The reconstruction of structures from projections is a problem that arises in many fields (for example in computerised tomography), and its mathematics is well known. This makes it possible to calculate the number of projections (the initial information) that allows a complete reconstruction of any given structure, and so it is also possible to define precisely what a reconstruction from incomplete information is. Such a reconstruction amounts to producing structures that belong to the object in question but for which there is insufficient initial information, and this is equivalent to saying that the reconstruction is producing a convergent increase of complexity. In the same way, to say that the phenotype of an organism is more complex than its genotype is equivalent to saying that any phenotype is reconstructed from a genotype which contains incomplete information.

The problem, of course, is to show that such reconstructions are possible, but this has been achieved by a particular class of iterative algorithms (Barbieri, 1974a). We have, therefore, mathematical models that allow us to simulate the problem of epigenesis in a meaningful way, and hopefully to understand the logic of its various steps. The second principle of semantic biology, in conclusion, is a new definition of epigenesis. It states that *epigenesis is a reconstruction from incomplete information*.

The third principle

The iterative algorithms that have been proposed for the reconstruction of structures from insufficient information differ from all other methods because they perform in parallel two distinct reconstructions: one for the structure matrix, and one for the so-called *memory matrix*, i.e. for a matrix where any convenient feature can be stored. This is why these algorithms are collectively referred to as the Memory Reconstruction Method (MRM).

With non-linear operations, for example, it is noticed that values appear at each iteration which are above the maximum or below the minimum. The space distribution of these "illegal" values is apparently random, but if they are recorded in the memory matrix, a new kind of information becomes available. It is seen that the illegal values are truly random only in some points, while in others they keep reappearing with regularity at each iteration. These last points are called *vortices* and, once recognised, they can be fixed and taken away from the number of the unknowns. This steadily decreases the unknowns, and when their number becomes equal to the number of equations a complete reconstruction can be performed in a straightforward way. The memory matrix, in other words, is a place where new information about the original structure appears, thus compensating for the incomplete information that was given at the beginning.

The memory space is the only space where such novel information can be found, and it follows therefore that any reconstruction from incomplete information is possible only if some kind of memory is used. In biology, this amounts to saying that any living system must contain two distinct types of structures: some have the visible role of the phenotype, while others act as depositories of information. The third principle of semantic biology, in short, states that *there cannot be a convergent increase of complexity without memory*. Or, in other words, *organic epigenesis requires organic memories*.

The fourth principle

The information that appears in the memory space cannot be transferred automatically to the structure space, and can be used only by employing specific conventions (the recognition of vortices in the memory matrix, for example, can be used only if a convention gives a meaning to the corresponding points of the structure matrix). This is another conclusion that leads to a universal principle, because it is necessarily valid for all systems.

New information can appear in a memory only if the memory space is truly independent from the structure space, because if they were linked (as real space and Fourier space, for example) one could only have the same information in different forms. Between two independent spaces, on the other hand, there is no necessary correspondence, and therefore a link can be established only by conventions, i.e. by the rules of a code.

This is the point where meaning enters the scene as a necessary entity, because the operation of establishing a correspondence between two independent worlds is equivalent to attaching a meaning to the structures of those worlds. Independent worlds, in other terms, can only be connected by codes, and if independent organic worlds do exist in life, then organic codes must also exist (the protein world and the nucleic acid world, for example, contribute to life only because there is a genetic code that builds a bridge between them).

The fourth principle of semantic biology, in short, states that *there cannot be a convergent increase of complexity without codes*. Or, in other words, *organic epigenesis requires organic codes*.

It may appear that only the fourth principle introduces the semantic dimension into biology, because it is only there that codes and meaning are explicitly mentioned, but this conclusion would be short-sighted. Organic codes and organic memories exist in life only because they are necessary to produce epigenetic systems, and so the fourth principle is dependent upon the idea that every living being is such a system (the first principle). In a similar way, any one of the above principles is a complement to the other three, and therefore all contribute to the building of semantic biology.

The first model

In 1981, the *Journal of Theoretical Biology* published "The ribotype theory on the origin of life", a paper which proposed two novel ideas: (1) an origin-of-life scenario based on ribosome-like particles, and (2) a theory of the cell as a system of three fundamental categories, more precisely as a system made of *genotype, ribotype* and *phenotype*. It is worth noticing that the term *ribotype* has later become fairly popular in the scientific literature, but has completely lost its original meaning. Now it is commonly used only to label RNA classes, and not to convey the idea that the ribotype is a true cell category, with the same "ontological" status as genotype and phenotype.

The origin-of-life scenario was instrumental for the new theory of the cell, because it led to the the conclusion that the ribotype had an evolutionary priority over genotype and phenotype. More precisely, the scenario described a precellular ribotype world (not to be confused with the RNA world) where some *ribosoids* could act as templates (ribogenotype), others as enzymes (ribophenotype), and others as polymerising ribosoids (ribotype) that were responsible for the growth and the *quasi-replication* of the ribonucleoprotein systems.

The first precellular systems were therefore made of three categories (ribogenotype–ribotype–ribophenotype) that evolved in different ways, the ribogenes being replaced by DNAs, and the ribozymes being dethroned by protein enzymes. In this way, the ribogenotype became a DNA genotype, and the ribophenotype turned into a protein phenotype, but the ancestral ribotype evolved without giving up the original function of making phenotypic products from genotypic instructions, even when quasi-replication evolved into exact replication. The precellular systems, in short, were based on three fundamental categories, and gave origin to cellular systems that have been based on equivalent categories ever since. Hence the idea that all cells have a genotype, a phenotype and a ribotype.

The ribotype cannot be given up because there is no DNA and no protein that can do the job of protein synthesis. Proteins and DNAs are two independent worlds, and only an organic code can build a bridge between them. The genetic code is that bridge, and that code is a quintessential RNA business (a correspondence between messenger RNAs and transfer RNAs). This concept can also be expressed in another way. As proteins are the seat of biological energy, and DNAs the seat of biological information, so RNAs are the seat of genetic coding, i.e. of biological *meaning*.

The first model of semantic biology, in conclusion, is the idea that *"The cell is a trinitary system made of genotype, ribotype and phenotype."* A more detailed version is *the semantic theory of the cell: "The cell is an epigenetic system made of three fundamental categories (genotype, ribotype and phenotype) which contains at least one organic memory (the genome) and one organic code (the genetic code)."*

The second model

The definition of epigenesis as a *reconstruction from incomplete information* suggests that embryonic development can be simulated (in a very abstract way) by the reconstruction of a super matrix made of a growing number of individual matrices, each of which would represent a cell. In this case, however, the reconstruction could be performed with two different strategies: one where the memory information is extracted only from individual memory matrices, and a second one where it is also extracted from a *collective* memory.

The biological equivalents of these strategies are two different kinds of embryonic development: one which exploits only cellular memories, and another which also makes use, from a certain point onwards, of a *supracellular* memory (the supracellular memory can exist only *from a certain point onwards*, because it is built by embryonic cells which must have already gone through a transformation phase).

The first kind of development (being continuous or single-phased) is an evolutionary precondition for the second one (which is two-phased or discontinuous), and this suggested that there might have been a transition from the first to the second developmental strategy in the history of life. Such a transition, incidentally, could well correspond to the Cambrian explosion, i.e. to the appearance of all known animal phyla in a geologically brief period of time.

Apart from the Cambrian explosion model, the interesting point is that the supracellular memory predicted by the reconstruction method does have the characteristics that are normally attributed to the *body plan*: they are both structures that appear from a certain point onwards in ontogenesis, and which function as depositories of supracellular information for the rest of the body's life. What has become known as the *phylotypic stage* of development, in other words, corresponds to the appearance of a structure which acts as a supracellular memory, and which can rightly be called *phylotype*, because it is characteristic of each phylum. And the phylotype is an intermediary between genotype and phenotype at the supracellular level, just as the ribotype is at the level of the single cell.

The second model of semantic biology, in conclusion, is the idea that *"An animal is a trinitary system made of genotype, phylotype and phenotype."* Another, more detailed, version of the model is *the semantic theory of embryonic development: "Embryonic development is a sequence of two distinct processes of reconstruction from incomplete information, each of which increases the complexity of the system in a convergent way. The first process builds the phylotypic body and is controlled by cells. The second leads to the individual body and is controlled not only at the cellular level but also at the supracellular level of the body plan."*

The third model

The scientific study of mental development has produced two outstanding discoveries. One is that there is an enormous gap between inputs and outputs (the so-called *poverty of the stimulus*), because children receive only very limited and erratic inputs of words in their learning period, and yet in the end they come up with a complete set of rules. The second is that children are predisposed to learn any language whatsoever, and so must develop, at some stage, a common *inborn mind*, a set of general rules that Noam Chomsky (1965) called *universal grammar*. So far, these discoveries have not been properly explained, probably because they have only been interpreted with *ad hoc* hypotheses. It may be worth noticing, therefore, that in the reference system of semantic biology they are accounted for in a very natural way.

The poverty of the stimulus is only another way of saying that mental structures are *reconstructions from incomplete information*, i.e. that they are the result of epigenetic processes. The universal grammar, on the other hand, is a structure that appears in human development *from a certain point onwards*, and which remains for the rest of the mind's life as a core deposit of information. As there is a phylotypic stage in embryonic development which is common to all members of a phylum, so there is a *specietypic stage* in mental development which is common to all members of our species. We can also say that as the body plan is the *phylotype* of an animal group, so the universal grammar is the *specietype* of mankind. According to semantic biology, in other words, mental development has the same fundamental logic of embryonic development, because both are reconstructions from incomplete information, and therefore both require memories and codes. The differences that divide them are mere by-products of the fact that organic structures and mental structures do not have identical physical substrates.

The third model of semantic biology, in conclusion, is the idea that *"The mind is a trinitary system made of mental genotype, mental specietype and mental phenotype."* Another version is *the semantic theory of mental development*: *"Mental development is a sequence of two distinct processes of reconstruction from incomplete information, each of which increases the complexity of the system in a convergent way. The first process builds the specietypic mind (the universal grammar), while the second leads to the individual mind."*

The fourth model

The idea that cultural evolution can teach us some deep truths about organic evolution has been dismissed by representatives of very different schools. The argument is that cultural evolution works with a Lamarckian mechanism, and therefore must have been produced by natural selection only after the arrival of nervous systems that were extravagant enough to start playing Lamarckian games.

Before such high-table reasoning, semantic biology can only take refuge in good old stubborn facts: Does the genetic code exist? Is it an organic code? Do signal transduction codes exist? Are they organic codes? Do splicing codes exist? And cell adhesion codes? And cytoskeleton codes, and compartment codes, and so on and on. If we agree that these are questions about nature-as-it-is (as opposed to nature-as-we-want-it-to-be), and if the answers are what the evidence tells us, then organic codes do exist. And if they exist, they had origins and histories. If organic meaning belongs to organic life, we must humbly accept that nature is just made that way. This is what semantic biology is about.

Such a view may help us to clarify at least two points of some weight. The first is Max Delbrück's question: *"How could mind arise from matter?"* (Delbrück, 1986). The answer from semantic biology is that organic life and mental life are both concerned with reconstructing structures from incomplete information, and so there is no vacuum between them. The materials are different, but the logic is the same. Nature could produce mental codes with the same craft with which she had been producing organic codes for 4 billion years.

The second point is about sudden changes in macroevolution. As long as the rules of an organic code evolve individually, not much seems to be happening, but when they are all in place and a new code emerges, something totally novel comes into existence. And that does explain how sudden changes of great magnitude could have taken place in the history of life.

Apart from these speculative detours, the basic issue is about the stuff of life, and the fourth model of semantic biology is merely the logical consequence of acknowledging the experimental reality of the organic codes. A particular version of that model is *the semantic theory of evolution: "The origin and the evolution of life took place by natural selection and by natural conventions. The great events of macroevolution have always been associated with the appearance of new organic codes."*

Conclusion

One day the above eight propositions will probably be regarded as a preliminary step, and that will be excellent news because it will mean that semantic biology has grown into a mature science, finally able to match the complexity of life. But that is a long way ahead. Today we have only just landed on an unexplored new continent, and we are in for many encounters with the unexpected. The natives seem friendly enough, though. The announcements of the *sugar code* (Gabius, 2000) and of the *histone code* (Strahl and Allis, 2000; Turner, 2000) prove that some pilgrim fathers have already reached a few nearby territories, and this is just the beginning. Our children will have the fortune of knowledge. We have before us the struggle and the thrill of discovery.

DEFINITIONS OF LIFE

(From Noam Lahav's *Biogenesis*, 1999; from Martino Rizzotti's *Defining Life*, 1996; and from personal communications by David Abel, Pietro Ramellini and Edward Trifonov, with permission).

Jean Baptiste LAMARCK (1802)

Life is an order or a state of things in the component parts of a body that makes organic movement possible and that effectively succeeds, as long as it persists, in opposing death.

Ludwig BÜCHNER (1855)

Spontaneous generation exists, and higher forms have gradually and slowly developed from previously existing lower forms, always determined by the state of the earth, but without immediate influence of a higher power.

Rudolf VIRCHOW (1855)

Life will always remain something apart, even if we should find out that it is mechanically aroused and propagated down to the minutest detail.

Ernst HAECKEL (1866)

Any detailed hypothesis concerning the origin of life must, as yet, be considered worthless, because up till now we have no satisfactory information concerning the extremely peculiar conditions which prevailed on the earth at the time when the first organisms developed.

Thomas Henry HUXLEY (1868)

The vital forces are molecular forces.

Justus von LIEBIG (1868)

We may only assume that life is just as old and just as eternal as matter itself ... Why should not organic life be thought of as present from the very beginning just as carbon and its compounds, or as the whole of uncreatable and indestructible matter in general?

J. BROWNING (1869)

There is no boundary line between organic and inorganic substances... Reasoning by analogy, I believe that we shall before long find it an equally difficult task to draw a distinction between the lowest forms of living matter and dead matter.

L.S. BEALE (1871)

Life is a power, a force or a property of a special and peculiar kind, temporarily influencing matter and its ordinary forces, but entirely different from, and in no way correlated with, any of these.

H.C. BASTIAN (1872)

Living things are peculiar aggregates of ordinary matter and of ordinary force which in their separate states do not possess the aggregates of qualities known as life.

Claude BERNARD (1878a)

Life is neither a principle nor a resultant. It is not a principle because this principle, in some way dormant or expectant, would be incapable of acting by itself. Life is not a resultant either, because the physicochemical conditions that govern its manifestation cannot give it any direction or any definite form ... None of these two factors, neither the directing principle of the phenomena nor the ensemble of the material conditions for its manifestation, can alone explain life. Their union is necessary. In consequence, life is to us a conflict.

Claude BERNARD (1878b)

If I had to define life in a single phrase ... I should say: life is creation.

Friedrich ENGELS (c. 1880)

No physiology is held to be scientific if it does not consider death an essential factor of life ... Life means dying.

Herbert SPENCER (1884)
> *The broadest and most complete definition of life will be "the continuous adjustment of internal to external relations".*

August WEISMANN (*c.* 1890)
> *The living organism has already been compared with a crystal, and the comparison is,* mutatis mutandis, *justifiable.*

Wilhelm PFEFFER (1897)
> *Even the best chemical knowledge of the bodies occurring in the protoplasm no more suffices for the explanation and understanding of the vital processes, than the most complete chemical knowledge of coal and iron suffices for the understanding of a steam engine.*

A.B. MACALLUM (1908)
> *When we seek to explain the origin of life, we do not require to postulate a highly complex organism ... as being the primal parent of all, but rather one which consists of a few molecules only and of such a size that it is beyond the limit of vision with the highest powers of the microscope.*

A. PUTTER (1923)
> *It is the particular manner of composition of the materials and processes, their spatial and temporal organisation which constitute what we call life.*

Alexander OPARIN (1924)
> *What are the characteristics of life? In the first place there is a definite structure or organisation. Then there is the ability of organisms to metabolise, to reproduce others like themselves, and also their response to stimulation.*

J.H. WOODGER (1929)
> *It does not seem necessary to stop at the word "life" because this term can be eliminated from the scientific vocabulary since it is an indefinable abstraction and we can get along perfectly well with "living organism" which is an entity which can be speculatively demonstrated.*

Ludwig von BERTALANFFY (1933)

A living organism is a system organised in a hierarchic order of many different parts, in which a great number of processes are so disposed that by means of their mutual relations, within wide limits with constant change of the materials and energies constituting the system, and also in spite of disturbances conditioned by external influences, the system is generated or remains in the state characteristic of it, or these processes lead to the production of similar systems.

Niels BOHR (1933)

The existence of life must be considered as an elementary fact that cannot be explained, but must be taken as a starting point in biology, in a similar way as the quantum of action, which appears as an irrational element from the point of view of classical physics, taken together with the existence of elementary particles, form the foundation of atomic physics.

Erwin SCHRÖDINGER (1944)

Life seems to be an orderly and lawful behaviour of matter, not based exclusively on its tendency to go from order to disorder, but based partly on existing order that is kept up.

J. ALEXANDER (1948)

The essential criteria of life are twofold: (1) the ability to direct chemical change by catalysis, and (2) the ability to reproduce by autocatalysis. The ability to undergo heritable catalysis changes is general, and is essential where there is competition between different types of living things, as has been the case in the evolution of plants and animals.

J. PERRETT (1952)

Life is a potentially self-perpetuating open system of linked organic reactions, catalysed stepwise and almost isothermally by complex and specific organic catalysts which are themselves produced by the system.

R.D. HOTCHKISS (1956)

Life is the repetitive production of ordered heterogeneity.

Norman HOROWITZ (1959)

I suggest that these three properties – mutability, self-duplication and heterocatalysis – comprise a necessary and sufficient definition of living matter.

Herman MULLER (1966)

It is alive any entity that has the properties of multiplication, variation and heredity.

John BERNAL (1967)

Life is a partial, continuous, progressive, multiform and conditionally interactive, self-realisation of the potentialities of atomic electron states.

Jacques MONOD (1970)

Living beings are teleonomic machines, self-constructing machines and self-reproducing machines. There are, in other words, three fundamental characteristics common to all living beings: teleonomy, autonomous morphogenesis and invariant reproduction.

Lila GATLIN (1972)

Life is a structural hierarchy of functioning units that has acquired through evolution the ability to store and process the information necessary for its own reproduction.

P. FONG (1973)

Life is made of three basic elements: matter, energy and information… Any element in life that is not matter and energy can be reduced to information.

Leslie ORGEL (1973)

Living beings are CITROENS (Complex Information-Transforming Reproducing Objects that Evolve by Natural Selection).

John MAYNARD SMITH (1975)

We regard as alive any population of entities which has the properties of multiplication, heredity and variation.

E. ARGYLE (1977)

Life on earth today is a highly degenerate process in that there are millions of different gene strings (species) that spell the one word "life".

Clair Edwin FOLSOME (1979)

Life is that property of matter that results in the coupled cycling of bioelements in aqueous solutions, ultimately driven by radiant energy to attain maximum complexity.

Manfred EIGEN (1981)

The most conspicuous attribute of biological organization is its complexity... The problem of the origin of life can be reduced to the question: "Is there a mechanism of which complexity can be generated in a regular, reproducible way?"

E.H. MERCER (1981)

The sole distinguishing feature, and therefore the defining characteristic, of a living system is that it is the transient material support of an organization with the property of survival.

E. HAUKIOJA (1982)

A living organism is defined as an open system which is able to maintain itself as an automaton ... The long-term functioning of automata is possible only if there exists an organisation building new automata.

Peter SCHUSTER (1984)

The uniqueness of life seemingly cannot be traced down to a single feature which is missing in the non-living world. It is the simultaneous presence of all the characteristic properties ... and eventually many more, that makes the essence of a biological system.

V. CSÁNYI and G. KAMPIS (1985)

It is suggested that replication – a copying process achieved by a special network of inter-relatedness of components and component-producing processes that produce the same network as that which produces them – characterises the living organism.

R. SATTLER (1986)

A living system is an open system that is self-replicating, self-regulating, and feeds on energy from the environment.

S. LIFSON (1987)

Just as wave–particle duality signifies microscopic systems, irreversibility means thermodynamic systems, and space-symmetry groups are typical of crystals, so do organisation and teleonomy signify animate matter.

Gerald EDELMAN (1988)

Animate objects are self-replicating systems containing a genetic code that undergoes mutation and whose variant individuals undergo natural selection.

Christopher LANGTON (1989)

Artificial Life can contribute to theoretical biology by locating life-as-we-know-it *within the larger picture of* life-as-it-could-be.

A. BELIN and J.D. FARMER (1992)

Life involves: (1) a pattern in spacetime (rather than a specific material object); (2) self-reproduction, in itself or in a related organism; (3) information-storage of a self-representation; (4) metabolism that converts matter/energy; (5) functional interactions with the environment; (6) interdependence of parts within the organism; (7) stability under perturbations of the environment; and (8) the ability to evolve.

Stuart KAUFFMAN (1993)

Life is an expected, collectively self-organised property of catalytic polymers.

A. de LOOF (1993)

Life is the ability to communicate.

Claus EMMECHE (1994)

Life itself is a computational phenomenon.

NASA's definition (Gerald JOYCE, 1994)

Life is a self-sustained chemical system capable of undergoing Darwinian evolution.

André BRACK (1996)

Life is a chemical system capable to replicate itself by autocatalysis and to make errors which gradually increase the efficiency of autocatalysis.

Sidney FOX (1996)

Life consists of proteinaceous bodies formed of one or more cells containing membranes that permit it to communicate with its environment via transfer of information by electrical impulse or chemical substance, and is capable of morphological evolution by self-organisation of precursors, and displays attributes of metabolism, growth, and reproduction. This definition embraces both protolife and modern life.

Tibor GÁNTI (1996)

At the cellular level the living systems are proliferating, program-controlled fluid chemical automatons, the fluid organisation of which are chemoton organisation. And life itself – at the cellular level – is nothing else but the operation of these systems.

Jesper HOFFMEYER (1996)

The basic unit of life is the sign, not the molecule.

Abir IGAMBERDIEV (1996)

Life is a self-organised and self-generating activity of open non-equilibrium systems determined by their internal semiotic structure.

Francisco VARELA (1996)

A physical system can be said to be living if it is able to transform external energy/matter into an internal process of self-maintenance and self-generation. This common sense, macroscopic definition, finds its equivalent at the cellular level in the notion of autopoiesis. This can be generalised to describe the general pattern for minimal life, including artificial life. In real life, the autopoietic network of reactions is under the control of nucleic acids and the corresponding proteins.

F. HUCHO and K. BUCHNER (1997)

Signal transduction is as fundamental a feature of life as metabolism and self-replication.

R.S. ROOT-BERNSTEIN and P.F. DILLON (1997)

Living organisms are systems characterised by being highly integrated through the process of organisation driven by molecular (and higher levels of) complementarity.

Kalevi KULL (1998)

An organism is a text to itself since it requires reading and re-representing its own structures for its existence, e.g. for growth and reparation. It also uses reading of its memory when functioning. This defines an organism as a self-reading text.

Hubert YOCKEY (2000)

The segregated, linear and digital character of the genetic message is an elementary fact and therefore essentially a definition of life. It is a gulf between living organisms and inanimate matter.

Thomas SEBEOK (2001)

Because there can be no semiosis without interpretability – surely life's cardinal propensity – semiosis presupposes the axiomatic identity of the semiosphere with the biosphere.

David ABEL (2002)

Life is a symphony of dynamic, highly integrated, algorithmic processes yielding homeostatic metabolism, development, growth and reproduction.

David KOSHLAND (2002)

If I were in ancient Greece, I would create a goddess of life whom I would call PICERAS, ... because there are seven fundamental principles (the Seven Pillars of Life) on which a living system is based: P (Program), I (Improvisation), C (Compartmentalization), E (Energy), R (Regeneration), A (Adaptability), and S (Seclusion).

Edward TRIFONOV (2002)

Life is an almost precise replication.

AFTERWORD

I am very grateful to Ward Cooper, who brought Barbieri's book to my attention. He knew of my interest, and somewhat of my experience, in the biology of reproduction and development, and particularly that I was searching for new explanations in the complexity sciences for just those coding and complexity-increase problems that Barbieri has illuminated. What came as a complete surprise to all three of us was that my teaching models for embryology, based solidly in experimental and observational (time-lapse filming) studies of embryos, should so precisely fit what Barbieri pleads for: "a third phase of development, and attention will turn again to the experiments".

   He did not know of them because he came into embryology by the route of studying chick embryos for his research on ribosome microcrystals, and textbook models were not in his search path. Because he had come in through ribosomes, the DNA-is-God-and-RNA-is-his-Prophet model of development so ably purveyed by Dawkins – genes are blueprints – was what he saw as biological orthodoxy. He was very critical of this simple-minded model, so apt for prokaryotes but so inadequate for real development where there is genuine increase in complexity. A typical silliness, representative of the DNA-blueprint paradigm, was Williamson's attempt to explain very similar larvae of not-very-closely-related crustacea by supposing that they had "caught" DNA instructions from each other (1992). What Barbieri had missed was the heterodox scholarship, from Conklin at the beginning of the twentieth century through Waddington and Zeeman, Dalcq and (early) Wolpert, with a few others that saw the genes as only part of the picture. Barbieri was clearly part of this thread, but did not really know it was there and didn't identify with it.

Because his connection, via ribosome function, was with rather unphilosophical biochemists, he was not appreciated as he would have been by the thoughtful evolution/developmental heterodoxy. This is quite strong now in the United Kingdom and Canada, is called affectionately "Evo-Devo" and has even pulled John Maynard Smith over from genetical orthodoxy. A prophet of this new orthodoxy is Brian Hall (1992). My own textbooks and review articles, and (after my philosopause*) my books with Stewart have presented this alternative view, based in observations and generalisations but without the benefit of a rigorous theoretical model. Such all-inclusive mathematical attempts as Thom (1975) were not immediately useful because not constraining, even as interpreted by Zeeman (1977). A rigorous model has now been provided by Barbieri, and I will attempt to set this observational and experimental material, and my generalisations of it, within the frame he has provided.

There were two distinct threads in the early twentieth century considerations of reproduction. The thread that led from the rediscovery of Mendel's work, through Morgan and then Fisher and Haldane, gave us population genetics and neo-Darwinism – heredity with chromosomes, and neo-Mendelism – cytogenetics with heredity. This, fuelled by fruit flies and peas, *Neurospora* and mice, seemed to address the important questions raised by Darwin's revolution in biology. It told us how differences were propagated through the generations. Dawkins carries this story today, and presents it very attractively in popular books.

But there was another thread, from von Baer and Haeckel through to Wilson and Conklin, that saw the increase of complexity in each development as the central issue: not how the heredity was distributed across the generations, but how that heredity contrived that the offspring developed *into* that next generation. Barbieri discusses this literature, but from the outside; I lived in it. Here organisms with available embryos, particularly sea urchins and marine worms, amphibians and the chick embryo (animals that didn't *have* genetics)

---

* A change of life that afflicts laboratory biologists of a particular age, and makes them write pretentious books that address the Big Questions.

were the fuel for experimental data. This second thread told us how similarities were re-generated in each generation – usually in excruciating detail, so embryology was regularly the most difficult subject in biology undergraduate syllabuses. Where Latin squares and chromosome dances were addressing the hereditary issues, Spemann's transplantations of "organisers" and Hamburger's stagings, Goodrich's homologies within and among vertebrate embryos, and Wilson's expositions of cleavage types among marine invertebrates, even de Beer's *Embryos and Ancestors* (1958) were addressing the problems of development. Students had a poor choice: they had to learn Fisher-type mathematics, particularly statistics, for the neo-Darwinism, or very complex developmental details of diverse organisms (usually as serial sections on microscope slides) for the embryology syllabus.

I was trained as an embryologist by 'Espinasse, and he made sure that I was as familiar with Dawidoff's *Embryologie des Invertebrés* and Lillie's *Development of the Chick* as with papers on feather development, my post-graduate topic. He encouraged my consideration of increase-of-complexity in the developing feather against the background of classical embryology. So, when I prepared the manuscript of my first text book, *Living Embryos* (1963), I began with a prejudice acquired from this early embryological literature. I had been impressed with, for example, Conklin (1918): "*In short the egg cytoplasm determines the early development and the sperm and egg nuclei control only later differentiations.* We are vertebrates because our mothers were vertebrates and produced eggs of the vertebrate pattern; but the color of our skin and hair and eyes, our sex, stature and mental peculiarities were determined by the sperm as well as by the egg from which we came." (italics in original) That idea, that the egg knew where it was going, and the early nuclei were just along for the ride, was amply justified by contemporary and later experiments. Enucleated eggs of some marine worms (*Chaetopterus*) nevertheless became kind-of-trochophores; the nuclear-transplant experiments of the Moores and others were consistent in showing that early development was controlled by the architecture, and probably by hereditary elements in eggs; these were later called informosomes and shown to be maternal messenger RNA (Davidson, 1976).

Up to the common pattern for their phylum, it seemed that developmental trajectories were under maternal control, zygotic nuclei were passive. In *Living Embryos* I went rather further than my evidence carried me, and invented the *phyletic stage* to which the varied eggs within a phylum converged, and from which they diverged towards their different adults. That book was very successful, not least because it introduced beautiful zebra-fish embryos into embryology teaching and made the subject much more accessible (partly because I simplified the concepts too far...). Because of the previous usage of "phyletic" by Haeckel for a different concept, Sander (1982) suggested we all change to *phylotypic stage*, and this has now become a central idea in animal development. The concept of the zootype has allowed us to model the divergence of phylotypic stages during evolution (Cohen, 1993). We can now begin to imagine the diversity of the Burgess Shale organisms as the evolutionary discovery and exploitation of this prototype of modern developments, as Barbieri has done on pp. 210–212.

This phylotypic idea corresponds, as Barbieri rather tentatively suggests, to his *memory* element. Because its prescription is out of phase with the reading of the code represented by the nuclear DNA, each embryo can rely on its egg architecture (with informosomes in specific array) as a "given" tape-player, and "play" its DNA tape. That the construction of that player is itself based in, derived from DNA instructions has been claimed, but this is simply not so: its derivation leads back to *interaction* between a previous player and DNA instructions. Stewart and I have presented thought-experiment models that show how two or more *totally different* true-breeding organisms could use precisely the same DNA. Or how one DNA sequence could provide the "blueprint" for several different morphologies. There is no mapping from DNA-sequence to morphology (Cohen and Stewart, 1994; Stewart and Cohen, 1997). This illustrates and emphasises just such a complicity between codes and memory as Barbieri has claimed must exist for the embryo to develop, to acquire complexity progressively and nearly determinately. The continuing unravelling of the detailed interactions between oocyte gradients and structures (products of genes such as *bicoid* and *Hb^{mat}*) and substances resulting from unwrapping – expression – of maternal mRNA products, with the central roles of maternal-effect homeotic genes, is consistent with this two-phase view.

There are other biological issues to which this two-step provenance of morphology have been related, but it took a long time before it was even "noticed". Von Baer's law claimed that embryology begins similarly, and diverges; and much has been made of Haeckel's illustrations of different vertebrate embryos that are simply three repeats of one woodcut. Garstang (1928) made much of the morphological relationships between some (neotenous) adults and other larvae, and produced some delightful verse on the subject. De Beer's rationalisation of the relationship between ontogenies and phylogenies, in *Embryos and Ancestors,* improved greatly on "the embryo climbs its own family tree" image, but barely mentioned that dissimilar eggs produced similar early embryologies before again diverging at later stages. Again, the beautiful extended treatment of this subject by Steven Gould in *Ontogeny and Phylogeny* (1977) was mostly related to post-phylotypic development and hardly addressed the convergence that so often precedes the divergence of closely related forms. As Barbieri suggested, there was no generally available explanation for the convergence/divergence (except mine, but he wasn't looking at textbooks). Those are the models that Barbieri found, and properly could discover no more than traces of two-step, memory-and-code, evidence.

There had of course been several attempts to relate these apparently different modes of development, prior to and after the phylotypic stage, to other biological constraints. Dalcq (1957) followed some of the early embryologists like Jenkinson and Kellicott in blaming the provision of yolk, from whose burden the embryo must escape to begin its trajectory toward adulthood. So the magic of gastrulation had to be entered from different directions according to how yolky the various kinds of blastomeres were: often, as in amphibia and lung-fish, just animal/vegetal differences, but in birds and mammals necessitating primitive streaks and other aberrations (see also Cohen, 1963). Others followed Wilson and Dawidoff in finding cleavage, especially odd forms like the variants of spiral cleavage in annelids and molluscs, as constraining these embryos into odd modes of gastrulation, after which they could enter organogenesis in more regular ways. Waddington, in his *Epigenetics of Birds* (1952), had a chapter where he compared birds with other vertebrates, in exactly that frame. In the 1960s many people discovered that the early, pre-gastrulation phase of development proceeded "normally" in antibiotics that prevented mRNA transcription (usually actinomycin D) or even protein production (puromycin), and the idea that zygote nuclear products were not produced during this early phase gained ground (Davidson, 1976) ("normally" is in quotes because it was later discovered that there *was* some transcription/translation in this phase, that was necessary for normal *later*, post-gastrulation development in many organisms). Because there was no overarching theoretical framework to which we could relate these differences of mechanism, such as Barbieri has now provided, we were at a loss to find other than mechanical ideas (like yolk distribution or cleavage patterns or successive transcription) to account for modes of development as evinced by *Baupläne* (compare Barbieri's pp. 196–198).

I had ideas that were caught on the two-step model, not because I had a theory of development like Barbieri's, but because it seemed so much easier to understand development if the tape recorder was there for the tape. (The idea that "information passes across the generations" had never convinced me: the next generation isn't *there* when it's sent; the information is supposed to "make" the next generation!) My *Reproduction* book (1979) used that two-step model with more assurance in its embryology chapters (post-actinomycin-D), and began to exploit it for phylogeny too: some larvae (like trochophores and plutei) seemed to be phylotypic stages in themselves, whereas others (caterpillars and zoeae – and human babies) seemed to be adapted for specific modes of life. My article introducing a symposium on *Maternal Effects in Development*, and a similar introduction to a symposium on *Metamorphosis*, used that theme to general approbation by the British embryological community. The phylotypic stage was when control of development passed from maternal to zygote provision of mRNA and protein structure, also when the egg structure complicated itself to the point where the nuclei were able to identify with their future fates. Wolpert's idea of the map and the book could begin to operate, as the embryo had attained its simplest map, which it would differentiate into sub-territories as it went on. The genesis of that first map had been an early concern of Wolpert's (Gustavson and Wolpert, 1961), but he later simply assumed it was there.

The two-step nature of development, some larvae *being* the phyletic stage and therefore with simple structure deriving from maternal organisation of the egg, led to a new picture of the origins of major phyla, standing on the shoulders of Garstang and de Beer (Cohen and Massey, 1983). Garstang's idea was that "dropping off" the later stages of life histories (he used "neoteny", but meant paedogenesis as well) would permit the larval structure to attain breeding ability; then it could function as the adult, breeding, after which new and different adult stages could be added. This did not sit well with 1980s models of development in which all of the "balanced" genome was beginning to be seen as operating throughout development, as Wolpert's "book". Massey and I thought the process could be much more radical. If adult stages were reeled back not just to the larval stage, but to the phylotypic stage, the *adult* structure would come to be much more dependent upon oocyte organisation in the ovary. Much morphology would be maternally controlled, as it is in nematodes – the discovery of so many maternal-effect genes, about a third, has been a major surprise from *Caenorhabditis*. Nematodes, we thought (even before the discovery of so many maternal-effect genes) were a good candidate group for this radical remodelling. When the organisation is so maternally based, the zygote genome is then freed to take new directions. It is not constrained to follow a developmental path, a conserved map and book, like its ancestors. In previous models the map had constrained the book, and the book the map (as handle and blade had each constrained the other in a Stone Age axe which had had several changes of each; it could never have evolved into a bow-and-arrow). This reduction to phylotypic larva seemed a usefully radical way to separate the complicity of map and book, and to free the lineage to redesign the map, and particularly to rewrite the book. Such Crustacea as *Sacculina* look as if they have done just this, with the adult phase resembling a fungus before it again produces the (necessarily typically barnacle) ovary, from whose eggs come regular-looking nauplii.

So what has Barbieri's model done to enable understanding, or at least explanation, of these two-step embryologies? I believe that it has given us a good theoretical basis that shows how the two-step ontogeny, with development passed from maternal to zygotic control at the phylotypic stage, seems to have been necessitated by the requirement for increasing complexity during development.

Prokaryotes, and some eukaryotes without development such as yeasts, do not have the problem. Their vegetative chemosynthetic processes simply continue, adding to the volume and surface of the organism, until there is enough for two organisms. Replication of the genome is fairly continuous, albeit with some "train-wrecks" that must be sorted by an aggregate of DNA-repair enzymes, in prokaryotes like *Escherichia coli*. Although, in simple non-cellular eukaryotes like yeasts, the biochemical emphasis changes before and during cell division, the organism doesn't become notably more complex, just bigger and with a spindle apparatus. The Barbieri problem does not arise. For eukaryote protozoans like *Paramecium* some of the problem is solved by having a parallel heredity for the cortex. But for cellular metazoans, that must make the next generation's organisation *from just one of their cells* (Cohen, 1985), the Barbieri issue is inescapable. Complexity must increase, from the simple one-cell organisation of the oocyte up to the breeding adult, in each generation. Clearly, it was the solution of the Barbieri problem that made metazoan life possible. The metaphyta ducked it, by and large, by growing as modules – modular growth does not increase complexity by anything like the extent that larvae-and-metamorphosis do! So did coelenterates.

So how do code and memory relate to the two-phase development of metazoans? The first and most obvious connection, that Barbieri himself made, is that the construction of the phylotypic stage in each embryology represents a "memory", what he calls the body plan supracellular memory. By using a tiny bit of code to arrange that the ovary makes oocytes of particular architecture from-and-within the structure of the adult, the next generation are started on their complexity-increasing way.

The simple way of presenting the next steps are as codes, in Barbieri's sense. As the oocyte begins its journey, possibly at ovulation but perhaps well before that, or even after fertilisation, its gene products must find their way to link surface receptors to chromosomal effectors. The first code is to prepare pre-proteins (and perhaps RNA lengths, but we know little about their travels yet) so that the egg's calcium channels, for example, are primed to allow a burst of Ca^{++} into the egg as the sperm penetrates. Other adaptors must link the sperm events to the nascent structures of the oocyte as fertilisation proceeds. Barbieri has shown us that each process, each interaction, assigns an arbitrary coding system, a series of adaptors, that prime receptors X and maps them to effectors Y (p. 110). In the simple case the leader strips on the pre-proteins get them to the proper places in the oocyte, just as receptor molecules are targetted to the surface of differentiated cells. But the oocyte has more potential than a normal differentiated cell. Gradients are set up, and the first transform of the *Bauplan* map appears. In some eggs, like those of a few tunicates, this is done in colour; in frog eggs, it results in the grey crescent; in mammal eggs it seems hardly to change the structure, but the later inner-cell-mass may come to contain only maternal mitochondria as a result of this early map. Cleavage ensues, and in a few eggs (some radial cleavers such as tunicates, and all the spiral cleavers) the cytoplasmic structure of the fertilised egg is faithfully retained in the mass of blastula cells (Cohen, 1963). Because of the assignment of "sticky" membrane in some parts of the embryo, tendencies to wander off in others, the blastula works toward the *Bauplan*; one of the best observational and experimental studies of this process was that of Gustavson and Wolpert on sea urchin embryos (1961). But once we know what to look for, the development of trochophores from spirally cleaving worm eggs, or Spemann's analysis of the potency of the vertebrate organisers, can be seen as multiple codes connecting effectors and receptors over the surfaces of the embryonic cells. The *Bauplan* – phylotypic stage – is not topologically complex, and a few well-controlled processes can achieve it from the oocyte architecture.

For some embryos, notably *Drosophila*, we have knowledge of the orchestration of new gene products, inhibiting and promoting transcription of others according to gradients, to edges, to the oocyte's placement of the germ-plasm organisms at one end of the egg, the homeotic gene products at the other. Peter Lawrence's book (1992) was well named *The Making of a Fly*, even though the maggot is the first product of this early embryology. Small peptides are the language of much of this early conversation; they are Barbieri's adaptors, connecting effectors to receptors in arbitrary ways (adaptors like tRNAs, pp. 97–98). Once the phylotypic stage is achieved, the cells are different in different places. It may be that in some embryos the old codes continue to act and complicate the situation as the nuclei react to the cytoplasm they find themselves in, and to the transduced signals on their cell surfaces. No wonder it took so long to establish the chemical nature of Spemann's organiser; it was a message, and the chemical nature of the messenger was unimportant provided that it linked a particular X to an important Y. Later, in mammals most obviously, blood-borne hormones do the same job: Xs like some of the anterior pituitary cells emit follicle-stimulating hormone to wake up Ys, perhaps the peri-oocyte cells in the ovary, and achieve what our literary friends would call "closure" as the oocytes set up their internal maps and codes as they prepare to become the phylotypic anchors again. It is vital that the phylotypic stage is a reliable simple map, achieved by a simple process. The codes that link processes in different members of the phylum can then proceed differently, one end being tied down, as it were, to phyletic precedent. Only with some of the Xs reliable can the Ys arrange to transform the lineage … only with some Ys reliable can Xs vary through time. It is the Stone Age axe again: each handle must fit the old blade, each blade must fit the old handle, each adaptor must fit into the coding system in such a way that a viable embryonic trajectory results. So evolution is bounded by lineage possibilities in just the ways we recognise.

Waddington was very close to seeing this point for the general embryology of metazoans. His ideas of chreodes, developmental trajectories that were resistant to change both from mutation and from environmental differences and that resulted in canalisation of development, sat upon just such complex-systems thinking as Barbieri's. His genetic-assimilation experiments, with cross-veinlessness and bithorax "phenocopies", precisely demolished the one-gene-one-character thinking of the neo-Mendelists ('Espinasse, 1956; Waddington, 1966). (The same simple-minded attempts at biochemical descriptions of embryological processes have moved Barbieri to criticise them today.) Waddington worried at the convergence-then-divergence of vertebrate development, knew that epigenetic argument could link this up (1952), but the actinomycin D results with pre-gastrulae had not yet been published. I leapt in to two-phase development, in *Living Embryos* (1963) because of Conklin, and then justified it by Waddington's epigenetic arguments. Then – after the actinomycin D experiments had shown the difference in transcription – I was much more confident in promoting it in the later 1970s (Cohen, 1977, 1979). However, now that we understand that the phylotypic stages are only a small topological distance from oocyte structures, and that this pair of structures is linked by codes, we can recognise these anchors of metazoan developments.

I took it as far as that. But Barbieri has now given it a theoretical basis as code and memory. Memory should not be taken literally, as the memory store in a computer; it is much more like the memory carried in a mind. This topological transformation of egg structure into phylotypic stage, on a basis of ancient zootype structure, obviously *is* the Barbieri memory carried by each lineage. He claimed it as such on pp. 212–215. But it is not a simple concept. It is part of two systems out-of-phase, the genetic code and the phylotypic read-out. It is as difficult to delineate or define as those constraints that reverberate down the lineage of the Stone Age axe, demonstrating that today's axe is what it is because of a memory of the ancient axe, instantiated as a constraint at each replacement!

Barbieri has proposed that his memory-code system is the *only* way that complexity can genuinely increase, and has the very persuasive reconstruction of three-dimensional structures, from two-dimensional projections with inadequate information, as a paradigm. While I have no doubt that this is *one* way of achieving controlled increase of complexity during embryonic development, I am not convinced that it is the only way. I am nearly convinced that it is *the* way terrestrial Metazoa have hit upon to achieve complex larvae and adults.

I am persuaded that the model, appropriately as semantic biology, fits the development of language in each human individual by "hard-wiring" a Chomsky structure, Barbieri's memory instantiated in each generation, followed by coding of input. This is a meta-model for Pinker's models, with the proviso that regular forms of speech are dealt with by memory-of-rules mechanisms, whereas irregulars are dealt with by memory-retrieval mechanisms (Pinker, 1999) – two rather different hard-wired Barbieri memories that he doesn't clearly distinguish in this book. It is clear to me that this theoretical framework enables us to *explain* increase-of-complexity in embryos as we never could before. But I am not persuaded that we *understand* increase-of-complexity in embryos yet. A further step is needed. I believe, and so does Ghiselin, that Barbieri has shown that his memory-code system is up to the task of making Metazoa. I am delighted that it fits the two-out-of-phase stages that I have been proposing since 1963, and that it gives my vision of development a theoretical basis. Probably there are critical experiments that could be performed within this framework, that could not be conceived before. I agree that this is a *sufficient* framework within which to explain embryology. But I am not persuaded that it is *the* necessary one.

Kauffman has written two books that are relevant here (1993, 2000). In *The Origin of Order* he set up a model system, N objects each connected by K connections to others, to parameterise this way of making pattern from non-pattern. Out of this came button-and-thread models, and a great variety of TV and radio programmes, and popular science books, that quote "*The Edge of Chaos*" as the amount of connectivity that makes things interesting: N of 100 needs a K of about 7 to achieve this. The details are not important for my point. The NK model was a very good model for showing that order could be – must be – crystallisable out of disorder or randomness, and that this order-generation could be predictable and could provide algorithmic learning in people who worked with it. But, and this was not emphasised in Kauffman's book just as it is not in Barbieri's, there are many *other* ways of getting order. Kauffman himself does not now think that the many life processes he explained by NK networks (numbers of genes and proteins, numbers of kinds of differentiated cells) are to be understood in that way – although they still can be, the explanation still works. He now chooses other ways, what he calls a more *general biology* (2000) as a new paradigm, and I find this much more persuasive. He writes of biology (and technology) expanding its phase space as it exploits the adjacent possible, and so on, a concept not conceivable within *NK* models. He never said, in *The Origin of Order*, that *NK* is the only way to get order from disorder, but that it was a useful example, perhaps a central one.

I believe that the same is true of Barbieri's model. I think that it is one good way to increase complexity predictably without enough prescriptive information. It explains brains whose wiring requires much more information than is available from the DNA sequence, and embryos like puffer-fish that seem to develop without a long enough knitting-pattern. It may be the way that Earth's Metazoa all do it. But I don't think it will necessarily be the way *all* life does it everywhere (Cohen and Stewart, 2002), that it is the only way to get progressively complex development in the general biology book (Kauffman, 2000). But it's the best model we've got so far.

November 2001   *Jack Cohen*

REFERENCES

Abel, D.L. 2002. Is life reducible to complexity? In G. Palyi, C. Zucchi, and L. Caglioti (eds.), *Fundamentals of Life*, pp. 57–72. Elsevier, Paris.

Alberts, B., Bray, D., Lewis, J., Raff, M., Roberts, K. and Watson, J.D. 1989. *Molecular Biology of the Cell.* (3rd edn 1994.) Garland Publishing Inc., New York.

Alexander, J. 1948. *Life, its Nature and Origin.* Reinhold Publishing Company, New York.

Altman, S. 1984. Aspects of biochemical catalysis. *Cell*, 36(2), 237–239.

Argyle, E. 1977. Chance and the origin of life. *Origins of Life*, 8, 287–298.

Barbieri, M. 1974a. Density Modulation reconstruction technique. In R.B. Marr (ed.), *Techniques of Three-Dimensional Reconstruction, Proceedings of an International Workshop held at Brookhaven National Laboratory, Upton, New York,* BNL 20425, pp. 139–141.

Barbieri, M. 1974b. A criterion to evaluate three-dimensional reconstructions from projections of unknown structures. *Journal of Theoretical Biology*, 48, 451–467.

Barbieri, M. 1981. The ribotype theory of the origin of life. *Journal of Theoretical Biology*, 91, 545–601.

Barbieri, M. 1985. *The Semantic Theory of Evolution.* Harwood Academic Publishers, London. (Italian edition, 1985, *La Teoria Semantica dell'Evoluzione*, Boringhieri, Torino.)

Barbieri, M. 1987. Co-information: a new concept in Theoretical Biology. *Rivista di Biologia–Biology Forum*, 80, 101–126.

Barbieri, M. 1998. The organic codes: the basic mechanism of macroevolution. *Rivista di Biologia–Biology Forum*, 91, 481–514.

Barbieri, M. 2001. *The Organic Codes: The Birth of Semantic Biology.* Pequod, Ancona. (Italian edition, 2000, *I Codici Organici. La nascita della biologia semantica,* Pequod, Ancona.)

Bastian, H.C. 1872. Quoted in Farley (1977), p. 124.

Beale, L.S. 1871. Quoted in Farley (1977), p. 89.

Belin, A. and Farmer, J.D. 1992. Artificial life: the coming evolution. In C.G. Langton, C. Taylor, J.D. Farmer and S. Rasmussen (eds.), *Artificial Life II*, pp. 815–838. Addison-Wesley, Redwood City, California.

Benner, S.A., Ellington, A.D. and Tauer, A. 1989. Modern metabolism as a palimpsest of the RNA world. *Proceedings of the National Academy of Sciences USA*, 86, 7054–7058.

Bernal, J. 1967. *The Origin of Life*. Weidenfeld & Nicolson, London.

Bernal, J.D. 1951. *The Physical Basis of Life*. Routledge & Kegan Paul, London.

Bernard, C. 1878a. *Lectures on the Phenomena of Life*. (Translated by H.E. Hoff, R. Guillemin and L. Guillemin, 1974.) Charles Thomas, Springfield, Illinois.

Bernard, C. 1878b. *An Introduction to the Study of Medicine*. (Translated by H.C. Greene, 1927.) Macmillan, New York.

Berridge, M. 1985. The molecular basis of communication within the cell. *Scientific American*, 253(4), 142–152.

Berridge, M.J. 1993. Inositol trisphosphate and calcium signalling. *Nature*, 361, 315–325.

Bohr, N. 1933. Light and life. *Nature*, 131, 457–459.

Bonner, J.T. 1988. *The Evolution of Complexity*. Princeton University Press, Princeton, New Jersey.

Boveri, T. 1904. *Ergebnisse über die Konstitution der chromatischen Substanz des Zellkerns*. Gustav Fisher, Jena.

Brack, A. 1996. Quoted in Rizzotti (1996), p. 9.

Browning, J. 1869. Quoted in Farley (1977), p. 75.

Büchner, L. 1855. Quoted in Farley (1977), p. 72.

Burks, A.W. 1970. *Essays on Cellular Automata*. University of Illinois Press, Urbana, Illinois.

Cahn, F. and Lubin, M. 1978. Inhibition of elongation steps of protein synthesis at reduced potassium concentrations. *Journal of Biological Chemistry*, 253 (21), 7798–7803.

Cairns-Smith, A.G. 1982. *Genetic Takeover and the Mineral Origins of Life*. Cambridge University Press, Cambridge, UK.

Calvin, W.H. 1996. *The Cerebral Code*. MIT Press, Cambridge, Massachusetts.

Carlile, M.J. 1980. From prokaryote to eukaryote: gains and losses. *Symposia of the Society for General Microbiology*, 38, 1–40.

Cavalier-Smith, T. 1987. The origin of eukaryote and archaebacterial cells. *Annals of the New York Academy of Sciences*, 503, 17–54.

Cech, T.R. 1986. RNA as an enzyme. *Scientific American*, Nov., 76–84.

Chomsky, N. 1965. *Aspects of the Theory of Syntax*. MIT Press, Cambridge, Massachusetts.

Chomsky, N. 1972. *Problems of Knowledge and Freedom*. Fontana, London.

Cohen, J. 1963. *Living Embryos*. (3rd edn with B. Massey 1980.) Pergamon Press, Oxford, UK.

Cohen, J. 1977. *Reproduction*. Butterworth, London.

Cohen, J. 1979. Introduction: maternal constraints on development. In D. R. Newth and M. Balls (eds.), *Maternal Effects on Development*, pp. 1–28. Cambridge University Press, Cambridge, UK.

Cohen, J. 1985. Metamorphosis: introduction, usages and evolution. In M. Balls and M.E. Bownes (eds.), *Metamorphosis*, pp. 1–19. Oxford University Press, Oxford, UK.

Cohen, J. 1993. Scientific correspondence: Development of the zootype. *Nature* 363, 307.

Cohen, J. and Massey, B.D. 1982. Larvae and the origins of major phyla. *Biological Journal of the Linnean Society,* 19, 321–8.

Cohen, J. and Stewart, I. 1994. *The Collapse of Chaos: Simple Laws in a Complex World*. Viking, New York.

Cohen, J. and Stewart, I. 2002. *Evolving the Alien: The New Xenoscience*. Ebury Press, London.

Conklin, E. G. 1918. *Heredity and Environment in the Development of Men*. Princeton University Press, Princeton, New Jersey.

Conway Morris, S. 1993. The fossil record and the early evolution of the Metazoa. *Nature*, 361, 219–225.

Crick, F.H.C. 1957. The structure of nucleic acids and their role in protein synthesis. *Biochemical Society Symposia*, 14, 25–26.

Crick, F.H.C. 1966. The genetic code – yesterday, today and tomorrow. *Cold Spring Harbor Symposia on Quantitative Biology*, 31, 3–9.

Crick, F.H.C. 1968. The origin of the genetic code. *Journal of Molecular Biology*, 38, 367–379.

Crick, F.H.C. 1970. Central Dogma of molecular biology. *Nature,* 227, 561–563.

Crowther, R.A., DeRosier, D.J. and Klug, A. 1970. The reconstruction of a three-dimensional structure from projections and its application to electron microscopy. *Proceedings of the Royal Society London, Series A, 317*, 319–340.

Csányi, V. and Kampis, G. 1985. Autogenesis: the evolution of replicative systems. *Journal of Theoretical Biology*, 114, 303–321.

Cuvier, G. 1828. *Le Règne animal distribué d'après son organisation.* Fortin, Paris.

Dalcq, A. M. 1957. *Introduction to General Embryology.* Oxford University Press, London.

Darwin, C. 1859. *On the Origin of Species by Means of Natural Selection.* John Murray, London.

Darwin, C. 1876. *Autobiography.* John Murray, London.

Davidson, E. H. 1976. *Gene Activity in Early Development,* 2nd edn. Academic Press, London.

Dawidoff, C. 1928. *Traité d'embryologie comparée des invertebrés.* Paris.

Dawkins, R. 1976. *The Selfish Gene.* Oxford University Press, Oxford, UK.

De Beer, G. R. 1958. *Embryos and Ancestors.* Clarendon Press, Oxford, UK.

De Loof, A. 1993. Schrödinger 50 years ago: "What is Life?". "The ability to communicate", a possible reply? *International Journal of Biochemistry*, 25, 1715–1721.

Delbrück, M. 1986. *Mind from Matter?* Blackwell, Palo Alto, California.

Descartes, R. 1637. *Discours de la Méthode.* Leiden, The Netherlands.

Dobzhansky, T. 1937. *Genetics and the Origin of Species.* Columbia University Press, New York.

Dover, G. 1982. Molecular drive: a cohesive mode of species evolution. *Nature*, 299, 111–117.

Dover, G. 2000. *Dear Mr Darwin: Letters on the Evolution of Life and Human Nature.* Weidenfeld & Nicholson, London.

Dyson, F. 1985. *Origins of Life*. Cambridge University Press, Cambridge, UK.

Edelman, G.M. 1988. *Topobiology*, p. 5. Basic Books, New York.

Eigen, M. 1971. Self-organization of matter and the evolution of biological macromolecules. *Naturwissenschaften*, 58, 465–523.

Eigen, M. 1981. Transfer-RNA: the early adaptor. *Naturwissenschaften*, 68, 217–228.

Eigen, M. and Schuster, P. 1977. The hypercycle: a principle of natural self-organization. *Naturwissenschaften*, 64, 541–565.

Eigen, M., Gardiner, W., Schuster, P. and Winkler-Oswatitsch, R. 1981. The origin of genetic information. *Scientific American*, 244 (4), 88–118.

Eldredge, N. and Gould, S.J. 1972. Punctuated equilibria: an alternative to phyletic gradualism. In T.J.M. Schopf (ed.), *Models in Paleobiology*, pp. 82–115. Freeman, Cooper & Co., San Francisco, California.

Emmeche, C. 1994. The computational notion of life. *Theoria-Segunda Epoca*, 9 (21), 1–30.

Engels, F. c. 1880. *Dialectic of Nature*. (Translated and edited by C.D. Dutt, 1940.) International Publishers, New York.

'Espinasse, P. G. 1956. On the logical geography of neo-Mendelism. *Mind, new series*, 257, 75–77.

Farley, J. 1977. *The Spontaneous Generation Controversy*. Johns Hopkins University Press, Baltimore, Maryland.

Fisher, R.A. 1930. *The Genetical Theory of Natural Selection*. Oxford University Press, Oxford, UK.

Flemming, W. 1882. *Zellsubstanz, Kern und Zelltheilung*. Vogel, Leipzig.

Folsome, C.E. 1979. *The Origin of Life*. W.H. Freeman, San Francisco, California.

Fong, P. 1973. Thermodynamic statistical theory of life: an outline. In A. Locker (ed.), *Biogenesis, Evolution, Homeostasis: A Symposium by Correspondence*, pp. 93–101. Springer-Verlag, Berlin.

Fortey, R. 1998. *Life: An Unauthorised Biography*. Flamingo, London.

Fox, S. 1996. Quoted in Rizzotti (1996), p. 67.

Fox, S.W. and Dose, K. 1972. *Molecular Evolution and the Origin of Life*. W.H. Freeman, San Francisco, California.

Gabius, H.-J. 2000. Biological information transfer beyond the genetic code: the sugar code. *Naturwissenschaften*, 87, 108–121.

Gánti, T. 1975. Organisation of chemical reactions into dividing and metabolizing units: the chemotons. *Biosystems*, 7, 189–195.

Gánti, T. 1996. Quoted in Rizzotti (1996), p. 103.

Garcia-Bellido, A., Lawrence, P.A. and Morata, G. 1979. Compartments in animal development. *Scientific American*, 241, 102–111.

Garstang, W. 1928. The origin and evolution of larval forms. *Report of the British Association for the Advancement of Science (D)*, 77–98.

Gatlin, L.L. 1972. *Information Theory and the Living System*. Columbia University Press, New York.

Gehring, W.J. 1987. Homeo boxes in the study of development. *Science*, 236, 1245–1252.

Gerard, D.R. 1958. Concepts in biology. *Behavioural Sciences*, 3, 92–215.

Gerhart, J. and Kirschner, M. 1997. *Cells, Embryos, and Evolution*. Blackwell Science, Oxford, UK.

Ghiselin, M.T. 1974. A radical solution to the species problem. *Systematic Zoology*, 23, 536–544.

Ghiselin, M.T. 1997. *Metaphysics and the Origin of Species*. State University of New York Press, Albany, New York.

Ghiselin, M.T. 2000. Cultures as supraorganismal wholes. *Perspectives in Ethology*, 13, 73–87.

Gilbert, W. 1986. The RNA world. *Nature*, 319, 618.

Gogarten, J.P., Kibak, H., Dittrich, P., Taiz, L., Bowman, E.J., Bowman, B.J., Manolson, M.F., Poole, R.J., Date, T., Oshima, T., Konishi, J., Denda, K., and Yoshida, M. 1989. Evolution of the vacuolar H^+-ATPase: implications for the origin of eukaryotes. *Proceedings of the National Academy of Sciences USA*, 86, 6661–6665.

Goodrich, E.S. 1958. *Studies on the Structure and Development of Vertebrates*, vols. 1 and 2. Dover, New York.

Gordon, R. 1999. The *Hierarchical Genome and Differentiation Waves: Novel Unification of Development, Genetics and Evolution*. World Scientific Press, Singapore.

Gordon, R., and Herman, G.T. 1974. Three-dimensional reconstruction from projections: a review of algorithms. *International Review of Cytology*, 38, 111–151.

Gordon, R., Bender, R. and Herman G.T. 1970. Algebraic Reconstruction Techniques (ART) for three-dimensional electron microscopy and X-ray photography. *Journal of Theoretical Biology*, 29, 471–481.

Gould, S. J. 1977. *Ontogeny and Phylogeny*. Harvard University Press, Cambridge, Massachusetts.

Gould, S.J. 1989. *Wonderful Life: The Burgess Shale and the Nature of History.* W.W. Norton, New York.

Gould, S.J. and Vrba, E.S. 1982. Exaptation: a missing term in the science of form. *Paleobiology*, 8, 4–15.

Gulick, J.T. 1888. Divergent evolution through cumulative segregation. *Journal of the Linnean Society London, Zoology*, 20, 189–274.

Gustavson, T. and Wolpert, L. 1961. The forces that shape the embryo. *Discovery, new series*, 22, 470–477.

Haeckel, E. 1866. *Generalle Morphologie der Organismen.* Georg Reimer, Berlin.

Haldane, J.B.S. 1929. The origin of life. *Rationalist Annual*, 3, 148–153.

Haldane, J.B.S. 1932. *The Causes of Evolution.* Longmans Green, London.

Hall, B. K. 1992. *Evolutionary Developmental Biology*. Chapman & Hall, London.

Hamburger, V. 1988. *The Heritage of Experimental Embryology: Hans Spemann and the Organizer.* Oxford University Press, New York.

Harold, F.M. 1986. *The Vital Force: A Study of Bioenergetics.* W.H. Freeman, New York.

Haukioja, E. 1982. Are individuals really subordinate to genes? A theory of living entities. *Journal of Theoretical Biology*, 99, 357–375.

Hertwig, O. 1894. *Zeit- und Streitfragen der Biologie.* Gustav Fisher, Jena.

Hoffmeyer, J. 1996. *Signs of Meaning in the Universe*. University of Indiana Press, Bloomington, Indiana.

Horowitz, N. 1959. On defining life. In F. Clark and R.L.M. Synge (eds.), *The Origin of Life on Earth,* pp. 106–107. Pergamon Press, Oxford, UK.

Hotchkiss, R.D. 1956. Quoted in Gerard (1958).

Hounsfield, G.N. 1972. *A Method of and an Apparatus for Examination of a Body by Radiation such as X or Gamma Radiation.* Patent Specification No. 1,283,915 Patent Office, London.

Hucho, F. and Buchner, K. 1997. Signal transduction and protein kinases: the long way from plasma membrane into the nucleus. *Naturwissenschaften*, 84, 281–290.

Huxley, T.H. 1868. Quoted in Farley (1977), p. 73.

Igamberdiev, A. 1996. Quoted in Rizzotti (1996), p. 129.

Iwabe, N., Kuma, K., Hasegawa, M., Osawa, S. and Miyata, T. 1989. Evolutionary relationship of archaebacteria, eubacteria, and eukaryotes inferred from phylogenetic trees of duplicated genes. *Proceedings of the National Academy of Sciences USA*, 86 (23), 9355–9359.

Jacob, F. 1982. *The Possible and the Actual*. Pantheon Books, New York.

Jenkinson, J.W. 1909. *Experimental Embryology*. Clarendon Press, Oxford, UK.

Johannsen, W. 1909. *Elemente der exacten Erblichkeitslehre*. Gustav Fisher, Jena.

Joyce, G.F. 1989. RNA evolution and the origins of life. *Nature*, 338, 217–224.

Joyce, G. 1994. Foreword. In D.W. Deamer and G.R. Fleischaker (eds.), *Origins of Life: The Central Concepts*, pp. xi–xii. Jones & Bartlet Publishers, Boston, Massachusetts.

Joyce, G.F., Schwartz, A.W., Orgel, L.E. and Miller, S.L. 1987. The case for an ancestral genetic system involving simple analogues of the nucleotides. *Proceedings of the National Academy of Sciences USA*, 84, 4398–4402.

Kauffman, S.A. 1986. Autocatalytic sets of proteins. *Journal of Theoretical Biology*, 119, 1–24.

Kauffman, S.A. 1993. *The Origin of Order: Self-Organization and Selection in Evolution*. Oxford University Press, New York.

Kauffman, S.A. 2000. *Investigations*. Oxford University Press, Oxford, UK.

Kellicott, W.E. 1913. *Outline of Chordate Development*. Holt, New York.

Kimura, M. 1968. Evolutionary rate at the molecular level. *Nature*, 217, 624–626.

Kimura, M. 1983. *The Neutral Theory of Molecular Evolution*. Cambridge University Press, Cambridge, UK.

Koshland, D.E. 2002. The Seven Pillars of Life. *Science*, 295, 2215–2216.

Kuhn, T.S. 1962. *The Structure of Scientific Revolutions*. University of Chicago Press, Chicago, Illinois.

Kull, K. 1998. Organisms as a self-reading text: anticipation and semiosis. *International Journal of Computing Anticipatory Systems*, 1, 93–104.

Lahav, N. 1999. *Biogenesis: Theories of Life's Origin*. Oxford University Press, New York.

Lamarck, J.B. 1802. *Recherches sur l'organisation des corps vivants*. Paris.

Lamarck, J.B. 1809. *Philosophie zoologique*. Paris. (English translation reprinted 1963, Hafner, New York.)

Langton, C.G. 1989. Artificial life. In C.G. Langton (ed.), *Artificial Life*, pp. 1–47. Addison-Wesley, Redwood City, California.

Lawrence, P. 1992. *The Making of a Fly*. Blackwell Scientific Publications, Oxford, UK.

Leopold, A.S. 1949. A *Sand County Almanac, with other Essays on Conservation*. Oxford University Press, New York.

Lewis, E.B. 1978. A gene complex controlling segmentation in *Drosophila*. *Nature*, 276, 563–570.

Lifson, S. 1987. Chemical selection, diversity, teleonomy and the second law of thermodynamics. *Biophysics and Biochemistry*, 26, 303–311.

Lillie, F.R. 1908. *The Development of the Chick*. Holt, New York.

Linnaeus (Carl von Linné). 1758. *Systema Naturae*, 10th edn. Stockholm.

Lorch, J. 1975. The charisma of crystals in biology. In Y. Elkana (ed.), *Interaction between Science and Philosophy*, pp. 445–461. Humanities Press, New York.

Lubin, M. 1964. Intracellular potassium and control of protein synthesis. *Federation Proceedings*, 23, 994–1001.

Macallum, A. 1908. Quoted in Farley (1977), p. 159.

Malthus, T.R. 1798. *An Essay on the Principle of Population, as It Affects the Future Improvement of Society*. J. Johnson, London.

Mange, D. and Sipper, M. 1998. Von Neumann's quintessential message: genotype + ribotype = phenotype. *Artificial Life*, 4, 225–227.

Maniatis, T. and Reed, R. 1987. The role of small nuclear ribonucleo-protein particles in pre-mRNA splicing. *Nature*, 325, 673–678.

Mantegna, R.N, Buldyrev, S.V., Goldberger, A.L., Havlin, S., Peng, C.K., Simons, M. and Stanley, H.E. 1994. Linguistic features of noncoding DNA sequences. *Physical Review Letters*, 73 (23), 3169–3172.

Maraldi, N.M. 1999. Trasduzione dei segnali a livello nucleare: evoluzione di un codice organico. *Systema Naturae–Annali di Biologia Teorica*, 2, 335–352.

Marchal, P. 1998. John von Neumann: the founding father of artificial life. *Artificial Life*, 4, 229–235.

Marchetti, C. 1980. Society as a learning system. *Technological Forecasting and Social Change*, 18, 267–282.

Margulis, L. 1970. *Origin of Eucaryotic Cells*. Yale University Press, New Haven, Connecticut.

Maynard Smith, J. 1975. *The Theory of Evolution*. Penguin Books, Harmondsworth, UK.

Maynard Smith, J. 1983. Models of evolution. *Proceedings of the Royal Society London, Series B*, 219, 315–325.

Maynard Smith, J. 1986. *The Problems of Biology*. Oxford University Press, Oxford, UK.

Maynard Smith, J. and Szathmáry E. 1995. *The Major Transitions in Evolution*. Oxford University Press, Oxford, UK.

Mayr, E. 1942. *Systematics and the Origin of Species*. Columbia University Press, New York.

Mayr, E. 1963. *Animal Species and Evolution*. Harvard University Press, Cambridge, Massachusetts.

Mayr, E. 1982. *The Growth of Biological Thought*. Harvard University Press, Cambridge, Massachusetts.

McClintock, B. 1951. Chromosome organization and genic expression. *Cold Spring Harbor Symposia on Quantitative Biology*, 16, 13–47.

McClintock, B. 1956. Controlling elements and the gene. *Cold Spring Harbor Symposia on Quantitative Biology*, 21, 197–216.

Mendel, G. 1866. Versuche über Pflanzen-hybriden. *Verhandlungen des naturforschenden Vereins Brünn*, 4, 3–57.

Mercer, E.H. 1981. *The Foundation of Biological Theory*. Wiley-Intersciences, New York.

Mereschowsky, C. 1910. Theorie der Zwei Pflanzenarten als Grundlage der Symbiogenesis, einer neuen Lehre der Entstehung der Organismen. *Biologisches Zentralblatt*, 30, 278–303, 321–347, 353–367.

Miller, S.L. 1953. A production of amino acids under possible primitive earth conditions. *Science*, 117, 528–529.

Miller, S.L. 1987. Which organic compounds could have occurred on the prebiotic earth? *Cold Spring Harbor Symposia on Quantitative Biology*, 52, 17–27.

Miller, S.L. and Orgel, L.E. 1974. *The Origin of Life on the Earth*. Prentice-Hall, Englewood Cliffs, New Jersey.

Monod, J. 1970. *Chance and Necessity*. A. Knopf, New York.

Moore, J.A. 1955. Abnormal combinations of nuclear and cytoplasmic systems in frogs and toads. *Advances in Genetics*, 7, 132–182.

Morgan, T.H. 1934. *Embryology and Genetics*. Columbia University Press, New York.

Muller, H.J. 1966. The gene material as the initiator and organizing basis of life. *American Naturalist*, 100, 493–517.

Murchison, R.I. 1854. *Siluria: The History of the Oldest Known Rocks containing Organic Remains*. John Murray, London.

Niesert, U., Harnasch, D. and Bresch, C. 1981. Origin of life between Scylla and Charybdis. *Journal of Molecular Evolution*, 17, 348–353.

Nitta, I., Kamada, Y., Noda, H., Ueda, T. and Watanabe, K. 1998. Reconstitution of peptide bond formation. *Science*, 281, 666–669.

Noller, H.F. 1991. Ribosomal RNA and translation. *Annual Review of Biochemistry*, 60, 191–228.

Noller, H.F., Hoffarth, V. and Zimniak, L. 1992. Unusual resistance of peptidyl transferase to protein extraction procedures. *Science*, 256, 1420–1424.

Nüsslein-Volhard, C. and Wieschaus, E. 1980. Mutations affecting segment number and polarity in *Drosophila*. *Nature*, 287, 795–801.

Oparin, A.I. 1924. *Proiskhozhdenie Zhizni*. Moskovskii Rabochii, Moscow. (English translation in J.D. Bernal (ed.), *The Origin of Life*, 1967, Weidenfeld & Nicholson, London.)

Oparin, A.I. 1957. *The Origin of Life on the Earth*. Academic Press, New York. ·

Orgel, L.E. 1973. *The Origins of Life.* John Wiley, New York.

Orgel, L.E. 1992. Molecular replication. *Nature,* 358, 203–209.

Ottolenghi, C. 1998. Some traces of hidden codes. *Rivista di Biologia–Biology Forum,* 91, 515–542.

Paley, W. 1803. *Natural Theology: Or, Evidences of the Existence and Attributes of the Deity, Collected from the Appearances of Nature,* 5th edn. Fauldner, London.

Perrett, J. 1952. Biochemistry and bacteria. *New Biology,* 12, 68–69.

Pfeffer, W. 1897. Quoted in von Bertalanffy (1933), p. 48.

Piattelli-Palmarini, M. 1989. Evolution, selection and cognition. *Cognition,* 31, 1–44.

Piattelli-Palmarini, M. 1999. Universalità e arbitrarietà del lessico. *Le Scienze, quaderni,* 108, 14–18.

Pinker, S. 1999. *Words and Rules: The Ingredients of Language.* Weidenfeld & Nicholson, London.

Poole, A.M., Jeffares, D.C. and Penny, D. 1998. The path from the RNA world. *Journal of Molecular Evolution,* 46, 1–17.

Portier, P. 1918. *Les Symbiotes.* Masson et Cie, Paris.

Putter, A. 1923. Quoted in von Bertalanffy (1933), p. 51.

Raff, R. 1996. *The Shape of Life.* University of Chicago Press, Chicago, Illinois.

Rasmussen, H. 1989. The cycling of calcium as an intracellular messenger. *Scientific American,* Oct., 44–51.

Redies, C. and Takeichi, M. 1996. Cadherine in the developing central nervous system: an adhesive code for segmental and functional subdivisions. *Developmental Biology,* 180, 413–423.

Rizzotti, M. 1996. *Defining Life: The Central Problem in Theoretical Biology.* University of Padua, Padua, Italy.

Root-Bernstein, R.S. and Dillon, P.F. 1997. Molecular complementarity, I: The complementarity theory of the origin and evolution of life. *Journal of Theoretical Biology,* 188, 447–479.

Sander, K. 1982. Oogenesis and embryonic pattern formation: known and missing links. In B. Goodwin and R. Whittle (eds.), *Ontogeny and Phylogeny,* pp. 137–170. Cambridge University Press, Cambridge, UK.

Sattler, R. 1986. *Bio-philosophy.* Springer-Verlag, Berlin.

Schimper, A.F.W. 1883. Über die Entwickelung der Chlorophyllkörner und Farbkorper. *Botanische Zeitung*, 41, 105–114.

Schleicher, A. 1869. *Darwinism Tested by the Science of Language*. John Camden Hotten, London.

Schrödinger, E. 1944. *What is Life?* Cambridge University Press, Cambridge, UK.

Schuster, P. 1984. Evolution between chemistry and biology. *Origins of Life*, 14, 3–14.

Sebeok, T.A. 1963. Communication among social bees; porpoises and sonar; man and dolphin. *Language*, 39, 448–466.

Sebeok, T.A. 2001. Biosemiotics: its roots, proliferation and prospects. *Semiotica*, 134, 61–78.

Shannon, C.E. 1948. A mathematical theory of information. *Bell System Technical Journal*, 27, 379–423, 623–656.

Simpson, G.G. 1944. *Tempo and Mode in Evolution*. Columbia University Press, New York.

Sipper, M. 1998. Fifty years of research on self-replication: an overview. *Artificial Life*, 4, 237–257.

Spemann, H. 1938. *Embryonic Development and Induction*. Yale University Press, New Haven, Connecticut.

Spencer, H. 1884. *The Principles of Biology*. D. Appleton & Co., New York.

Spiegelman, S. 1967. An *in vitro* analysis of a replicating molecule. *American Scientist*, 55, 3–68.

Sporn, M.B. and Roberts, A.B. 1988. Peptide growth factors are multifunctional. *Nature*, 332, 217–219.

Steitz, J.A. 1988. "Snurps". *Scientific American*, 258, 56–63.

Stewart, I. and Cohen, J. 1997. *Figments of Reality: The Origins of the Curious Mind*. Cambridge University Press, Cambridge, UK.

Strahl, B.D. and Allis, C.D. 2000. The language of covalent histone modifications. *Nature*, 403, 41–45.

Strohman, R.C. 1997. The coming Kuhnian revolution in biology. *Nature Biotechnology*, 15, 194–200.

Sutherland, E.W. 1972. Studies on the mechanism of hormone action. *Science*, 177, 401–408.

Sutton, W.S. 1903. The chromosomes in heredity. *Biology Bulletin*, 4, 231–251.

Szathmáry, E. 1989. The emergence, maintenance and transitions of the earliest evolutionary units. *Oxford Surveys in Evolutionary Biology*, 6, 169–205.

Tempesti, G., Mange, D. and Stauffer, A. 1998. Self-replicating and self-repairing multicellular automata. *Artificial Life*, 4, 259–282.

Thom, R. 1975. *Structural Stability and Morphogenesis.* Benjamin, Reading, Massachusetts.

Trifonov, E.N. 1988. Codes of nucleotide sequences. *Mathematical Biosciences*, 90, 507–517.

Trifonov, E.N. 1989. The multiple codes of nucleotide sequences. *Bulletin of Mathematical Biology*, 51, 417–432.

Trifonov, E.N. 1999. Elucidating sequence codes: three codes for evolution. *Annals of the New York Academy of Sciences*, 870, 330–338.

Trifonov, E.N. 2002. Personal communication.

Turner, B.M. 2000. Histone acetylation and an epigenetic code. *Bioessays*, 22, 836–845.

Van Beneden, E. 1883. Recherches sur la maturation de l'oeuf et la fécondation et la division cellulaire. *Archives de Biologie*, 4, 265–640.

Varela, F.J. 1996. In Rizzotti (1966), p. 149.

Varela, F.J., Maturana, H.R. and Uribe, R. (1974). Organization of living systems, its characterization and a model. *Biosystems*, 5, 187–196.

Virchow, R. 1855. Quoted in Farley (1977), p. 54.

Von Baer, K.E. 1828. *Entwicklungsgeschichte der Thiere.* Bornträger, Königsberg.

Von Bertalanffy, L. 1933. *Modern Theories of Development: An Introduction to Theoretical Biology.* Harper & Brothers, New York.

Von Liebig, J. 1868. Quoted in Engels (*c.* 1880), p. 190.

Von Neumann, J. 1966. *Theory of Self-Reproducing Automata.* University of Illinois Press, Urbana, Illinois.

Wächtershäuser, G. 1988. Before enzymes and templates: theory of surface metabolism. *Microbiological Reviews*, 52, 452–484.

Wächtershäuser, G. 1992. Groundworks for an evolutionary biochemistry: the iron–sulphur world. *Progress in Biophysics and Molecular Biology*, 58, 85–201.

Waddington, C. H. 1952. *The Epigenetics of Birds.* Cambridge University Press, Cambridge, UK.

Waddington, C. H. 1966. *Principles of Development and Differentiation.* Macmillan, New York.

Wallin, J.E. 1927. *Symbionticism and the Origin of Species.* Williams & Wilkins, Baltimore, Maryland.

Watson, J.D. and Crick, F.H.C. 1953. A structure for deoxyribose nucleic acid. *Nature*, 171, 737–738.

Weismann, A. 1885. *Die Kontinuität des Keimplasmas als Grundlage einer Theorie der Vererbung.* Gustav Fisher, Jena.

Weismann, A. *c.* 1890. Quoted in Lorch (1975).

Weismann, A. 1893. *The Germ Plasm.* Charles Scribner's Sons, New York.

White, H.B. 1976. Coenzymes as fossils of an earlier metabolic age. *Journal of Molecular Evolution*, 7, 101–104.

Williamson, D. I. 1992. *Larvae and Evolution: Toward a New Zoology.* Chapman & Hall, London.

Wilson, E.B. 1928. *The Cell in Development and Heredity.* Macmillan, New York.

Woese, C.R. 1970. Molecular mechanism of translation: a reciprocating ratchet mechanism. *Nature*, 226, 817–820.

Woese, C.R. 1981. Archaebacteria. *Scientific American*, 244 (6), 94–106.

Woese, C.R. and Fox, G.E. 1977a. Phylogenetic structure of the prokaryotic domain: the primary kingdoms. *Proceedings of the National Academy of Sciences USA*, 74, 5088–5090.

Woese, C.R. and Fox, G.E. 1977b. The concept of cellular evolution. *Journal of Molecular Evolution*, 10, 1–6.

Wolpert, L. 1969. Positional information and the spatial pattern of cellular differentiation. *Journal of Theoretical Biology*, 25, 1–47.

Woodger, J.H. 1929. *Biological Principles: A Critical Study.* Kegan Paul, French, Trubner & Co., London.

Wright, S. 1921. Systems of mating. *Genetics*, 6, 111–178.

Wright, S. 1931. Evolution in Mendelian populations. *Genetics*, 16, 97–159.

Yockey, H.P. 2000. Origin of life on earth and Shannon's theory of communication. *Computers and Chemistry*, 24 (1), 105–123.

Zeeman, E.C. 1977. *Catastrophe Theory: Collected Papers.* Addison–Wesley, New York.

INDEX

Printed in the United States
By Bookmasters